NATIVE USE OF FISH
IN HAWAII

NATIVE USE OF FISH IN HAWAII

MARGARET TITCOMB

with the collaboration of
MARY KAWENA PUKUI

The University Press of Hawaii
Honolulu

Originally published in 1952 as Memoir 29 of the Polynesian Society,
Wellington, New Zealand

Second edition 1972; paperback 1977

Library of Congress Catalog Card Number 78–153932
ISBN 0–8248–0592–5
Manufactured in the United States of America

Contents

TWENTY years ago when Margaret Titcomb was finishing her manuscript for this book there was little concern that the oceans of the earth might be endangered. Without doubt contamination of the seas was occurring, but there was no real public awareness—no sense of threat. A deleterious mercury content had not yet been detected in the great billfish and tunas. There were fewer oil despoliations, and the dumping of chemicals and radioactive wastes had not yet reached a level which would, in the next two decades, cause great submarine areas to be laid waste. Nor had commercial fisheries yet developed their omnivorous technical expertise to efficiency levels which, unless curbed, could wipe out whole species.

There was another difference. Then, there were "a few Hawaiian fishermen still living who were trained in the Hawaiian knowledge" (p. 54), and who could supplement the prodigious fund of information of Mary Kawena Pukui, with whose collaboration this book was written. Now, as the Hawaiian fishermen pass from the scene, there goes with them much of the ancient lore of the sea.

There is today an encouraging awareness on the part of scientists and the public at large—even among public officials—in respect to our threatened and diminishing resources. And it is to be hoped that this general awareness will be translated into restoration measures and programs. Perhaps contributions of the kind provided by this work can be a positive influence. That it might well be would please its author, for she has long been an ardent conservationist. But there is another reason for welcoming the renewed availability of this book. It is packed with information. And it will please both those primarily interested in fish and those devoted to learning about ancient Hawaiian culture.

ROLAND W. FORCE
Director, Bishop Museum
November 9, 1971

AUTHOR'S NOTE

THE toned illustrations in this volume were reproduced
from watercolors by the British artist Robert C. Barnfield
who lived in Hawaii for two years in the early 1890s. Barnfield kept diaries of two Pacific stays—at the Line Islands
in 1885 and in Hawaii later. He also visited Samoa.

The diaries are good reading in themselves. In Hawaii
he took great interest in the news of the day and filled pages
of his diary with newspaper accounts of Kalakaua's last
farewell to Hawaii and the events after Kalakaua's death.
Among Barnfield's many friends and acquaintances were
the artists Allen Hutchinson and W. F. Cogswell. He mentioned giving lessons in art to a few women in Honolulu,
but he said very little about his own work and nowhere
refered to his watercolors of fishes. One entry in October
1891 mentions the sending of "8 sheets of my fish drawings"
to the curator of the British Museum. When they were
returned in December by A. Günther, the artist expressed
no regret. Perhaps Günther wanted scientific drawings.

During his years in Hawaii, Barnfield was evidently in
very poor health, bouts with asthma and bronchitis sometimes lasting a week at a time. Occasionally days were
spent just recovering from devastating attacks. The abrupt
ending of his diary leaves one with the foreboding that that
day was nearly his last. The one sentence for March 15,
1893, is, "Handed over Mr. Emmeluth's child's portrait
which proves entirely satisfactory." He died two months
later in Honolulu, at the age of 38.

Barnfield's watercolors are a fine addition to our knowledge of tropical fishes, for the renderings are faithfully done
and the colors are exquisite. Very welcome are the artist's

notes accompanying the paintings. They describe variations in color of the fishes, where and when each was caught, and peculiarities of their habits.

Finding this collection of watercolors was a great stroke of luck. In London, in 1966, I visited the famous old bookshop of Maggs Brothers, on Berkeley Square. After greetings, I asked whether anything new in the way of old books had come in lately, and Mr. Maggs replied, "Well, there is a box down cellar that you can look at if you like." So down we trudged through a narrow lane, full bookshelves on either side, and at the extreme end (I think the shop must own cellars of adjacent houses!) there was an ordinary packing box, already opened. Mr. Maggs left me and I searched through, seeing copies of old friends. At the bottom was a worn album case that may have adorned a "parlor" table for years. I groaned, "No doubt a lot of photos of people I don't know!" But I untied the knot, and found to my great delight the lovely watercolors by Barnfield. His name was familiar to me; he had copied some Webber drawings which are in Bishop Museum. His heart had not been in that job, but it *had* been in depicting fish. I looked through the lot, and bounced upstairs. "There is something here that we *must* have. I myself have written about Hawaiian fish and these illustrate them beautifully." This technique was not that of a shrewd buyer, but Mr. Maggs did not take advantage of my bubbling enthusiasm, and the Bishop Museum was able to purchase the lot.

The paintings are precious, and to have some included here is a great joy to this author, as is also the interest shown by The University Press of Hawaii in issuing this work again. *Mahalo nui loa.*

NATIVE USE OF FISH
IN HAWAII

Introduction

FISH, including shellfish, were the main protein-giving elements of the Hawaiian diet. Pig, dog, chicken and wild birds furnished some additional proteins but the comparatively small supply marked them more for the chiefs' than the commoners' use. Daily life was one of fishing and cultivating the plantations. Fishing required a search of the sea, from the areas within the reefs to the sea scarcely within sight of land. By salting, drying, impounding, the supply was made somewhat independent of weather conditions. Care was taken to avoid waste.

The sea was a great reservoir of food for the Hawaiians and they were fond of a wide variety; probably everything edible was consumed. "There is no animal food which a Sandwich Islander esteems so much as fish," said a visitor in 1834 (6, vol. 1, page 215). A catch was portioned out to all within the 'ohana, or related community. When there was food no one went hungry. When supplies were abundant there was hearty indulgence in the joy of eating; when scarce, endurance was eased by the knowledge that effort would bring further supplies, except for the calamity of war or the occasional periods of long stormy weather.

Chiefs became epicurean in their taste, demanding rarities, or regal service, such as the supplying of live fish from far places. Priests prescribed certain fishes as acceptable to the gods, sometimes a fish was the essential object to offer the gods as well as eat after a period of sickness, and fish were used in some other ceremonies.

Certain sea creatures, most commonly sharks, sometimes became 'aumakua (personal gods) and were fed with regularity and recognized as individuals. Legends and

1

chants contain some characters that change at will to sea creatures, and there are numerous incidents in Hawaiian oral literature that reflect intimate knowledge of fish, their characteristics, habits and domain. The sea and its creatures were almost as well known as the life and attributes of land areas. In property divisions the ideal unit extended from the mountain top down to the shore and beyond into the sea, stopping only at the reef, or about a mile out, if there were no reef. In the Hawaiian mind, there was a balance between sea and land. Most of the important land creatures had counterparts in the sea which enabled priests to accept the sea counterpart for an offering to the gods if the land creature were unobtainable.

The sea was the great highway between shore localities, and between islands. Though there were trails, some paved with flat stones, the easiest way of getting from one shore area to another was by water. Therefore the bulk of the population preferred to live along the shores. Like other Polynesians, Hawaiians were able swimmers, navigators and seamen. Several early voyagers commented on their being almost amphibious.

The preparation of this paper on use of fish by Hawaiians, use dictated by attributes of their own culture, was started at the suggestion of Dr. E. S. C. Handy, Ethnologist, Bernice P. Bishop Museum, several years ago. The offering would have been small indeed without the help at every turn of Mrs. Mary Kawena Pukui, Hawaiian translator at Bishop Museum. Much material was set down from her own store and from that of her mother, Mrs. Pa'ahana Wiggin, of Ka'u, Island of Hawaii. Mrs. Pukui also segregated material as she translated legends at Bishop Museum, and all statements have been checked against her knowledge.

I am greatly indebted to the following kindly persons: to Dr. E. W. Gudger, Department of Fishes, American Museum of Natural History, for editorial comments and Mr. John T. Nichols, Department of Fishes, American Museum of Natural History, for encouragement, to Dr. Leonard P. Schultz, Curator of Fishes, U.S. National Museum, and to Dr. David Bonnet, Zoologist, formerly of the University of Hawaii, for corrections and criticisms of the descriptive list of fishes, to Mr. Vernon Brock, of the staff of the Board of Agriculture and Foresty, Honolulu, for general biological criticism, and especially for corrections suggested for the sketches, to Dr. Douglas Oliver, Ethnologist, for criticism of the whole plan and scope, to Dr. Kenneth P. Emory, Ethnologist, Bishop Museum, for criticism of the Hawaiian lore, to Dr. John Embree, for editorial and ethnological criticism, and to one ichthyologist in the Armed Forces who read with care the statement on Hawaiian nomenclature, and who prefers to remain unacknowledged. The offered corrections were made; not all suggestions could be followed, therefore all errors and omissions are attributable to the author alone.

Many Hawaiian informants and some others gave information and a list of their names follows. I hope those who are still living will be glad to see their knowledge recorded: Lily Akana, Hilo, Hawaii; Charles Alona, Waimanalo, Oahu; Hamana Kalili, Laie, Oahu; Sarah Jacobs, Hana, Maui; Edward Kaauwana Aukai, Kualoa, Oahu; Kalokuokamaile, Kona, Hawaii; Keliikipi Kanakaole, Ka'u, Hawaii; William K. Kinney, of Kauai and Maui; Joseph Kukea, Oahu; Lizzie Maka, Manoa, Oahu; Makahonu Naumu, Waimea, Kauai; Walanika Paka, Manoa, Oahu; William Watson, Kaneohe Bay, Oahu; Elizabeth Lahilahi Webb, Ewa and Honolulu, Oahu; Helen Kuehu, Hanalei, Kauai; George Manuia Galbraith, Kaalaea, Oahu; Joseph Kawelo, Kaalaea, Oahu; Nakuina, Honolulu, Oahu; William G. Anderson, Kaneohe Bay, Oahu; Yoshio Kondo, Honolulu, Oahu; E. Y. Hosaka, Honolulu; Thomas Maunupau, Honolulu and Kona, Hawaii, and a few others noted in the text.

The illustrations are rough sketches without scientific detail; no regard has been paid to the dorsal spine count. Inspiration for most of them is contained in Jordan and Evermann's work on Hawaiian fishes, a few are drawn from casts in Bishop Museum, a few from elsewhere.

PROCURING FISH

Fishing was one of the constant, necessary occupations. Everyone knew how to obtain fish by various techniques. The slave, the commoner, the lesser chiefs, the high chiefs, men, women and children got food from the sea by their own efforts. For some it was a duty, for most it was also a pleasure, for the chiefs it was a favourite sport. Children played about the shores and took what they pleased and could get from the shore pools, and shallow reef areas, and ate it when and as they pleased, raw or cooked. When old enough to follow their elders they learned by imitation how to get small fish and shellfish and *limu* (seaweed) from the sheltered waters, and later how to fish in deeper waters.

Every day saw many people, women in the majority, out on the reefs for hours, searching, collecting all that was edible and desirable. Calabashes tied to their persons floated along and held the catch. Doubtless the women made a merry social time of it too. To women belonged also the larger part of the task of gathering fish and shellfish from the mountain pools and streams, the *'o'opu* (gobies) and little shrimps (*'opae*) and other varieties. This collecting was done chiefly by feeling with the hands, poking with a stick, turning over stones and logs, with a net ready to catch the animals that darted out from cover. In times of freshet, men doubtless did the work of building a platform across a stream just under high water level to sluice off the less muddy waters where the *'o'opu* took refuge from the silt washed into the stream. These waters were led off on to a plain and the fish were stranded on the porous soil, and easily picked up (4, p. 8).

The chiefs were as fond of ocean fishing as commoners and went fishing as a pastime, either alone or with a few companions, or grandly with a large number of retainers, and with the *po'o lawai'a* (head fisherman). Meares (68a, p. 353) notes an occasion when Kamehameha I, the most powerful chief in Hawaii, came in from a fishing party. When the first group of missionaries came to Hawaii and were trying to persuade Kamehameha II to let them settle at Oahu, Ka'ahumanu, the widow of Kamehameha I, joined the conference, "having come in from a successful fishing expedition in a double sailing canoe" (43, p. 119). Some kinds of fishing called for such a lot of gear that only chiefs or professional fishermen could use those methods. There were nets many fathoms deep, and a greater number of fathoms long, for surrounding a school of fish at sea. There was *niuhi* (man-eater shark) fishing that required vast amounts of bait as lure. Mrs. Beckley (4, p. 19) called it the game of kings.

Fishing as a profession belonged to the *po'o lawai'a* and his company of apprentices. He went fishing with the chief, when that was the chief's pleasure. He fished at the order of the chief, transmitted through the chief's *kahu* (steward), or he went fishing when he wished to do so himself. Fishing was his life's occupation. He might be a chief himself, of lower rank than the high chief under whom he lived, or he might be a commoner. He was, at any rate, a man of extensive knowledge, and highly honoured. Most of his knowledge was handed down to him from an older relative or a friend. Such teachers chose their legatees with great care, and it was a pupil's duty to transmit his learning, augmented by his own experience, to his own chosen pupils. His knowledge comprised the techniques of manufacture and use of apparatus needed, though it was usually made for him by other craftsmen, the methods of capture, habitats of the various fish, seasons of their spawning, and of their coming and going if they were roaming fish that moved in schools, and their particular peculiarities of response to attempts at capture.

Most fishing expeditions started before dawn, the fishermen getting up and assembling silently, without speech that might offend the gods, and spoil the luck. The *po'o lawai'a* had to know how to judge the weather, how to divine the meaning of omens in dreams and in the clouds, and how to recognise the stars as indicators of time and direction, bird flights as indicators of schools of fish—birds being rivals of fishermen in catching and consuming fish. He had to be en rapport with the gods of fishing and his own personal gods, and avoid the enmity and therefore the curses of his fellow-men. He had to know how to manipulate his canoe, how to right it at sea, as did all Hawaiians. And he had to choose and train and manage his assistants, for most fishing beyond the reef was done by fishermen in concert. On shore there might be one more member of the party, the *kilo*, or watcher. Posted on a high point of land, this man watched for the expected schools of fish. Like a band leader, he had his own individual manner of signalling. Some used long pieces of bamboo, and some had other aids to make themselves more easily seen, and some used only their arms. By the *kilo's* motions he steered the fleet of canoes around the school. When he ceased signalling and sat down it was equivalent to saying, "You've got it." The *kilo* went to sea very little, but he did go when *mālolo* (flying fish) was the object of the expedition for a school of *mālolo* is often beyond signalling distance from shore. Mrs. Beckley says (4, p. 18) that the success of surrounding a school was entirely up to the *kilo*. It has been assumed by some that Hawaiians did not get far from land and did no deep-sea fishing. Beckley (4, p. 10) states that fishing canoes sometimes went " so far out from land as to be entirely out of sight of the low lands and mountain slopes and took their bearing . . . from the positions of the different mountain peaks . . ."

The full story of methods and customs in fishing is extensive. It is certain that everyone ate fish, and fishing was a constant occupation. Kelly (50, p. 9) says, " In my opinion, no people ever lived who had a more intimate knowledge of fish and their habits, and knew so well how to catch them as the Hawaiians . . ."

PONDS

Though walled traps of one form or another were (and many still are) used all over the Pacific, Hawaiians developed the use of ponds to a greater extent than any other people. There were two kinds—salt water ponds (*kua pā*: walled in back) and fresh water ponds (*haku one*: heaps of sand). The building of a fishpond was a laborious undertaking, and the credit for having one constructed was so great that it is still known which chiefs were the builders of some of them. Traditionally, the stone supply for the construction was assembled by establishing a line of men between the supply and the site of the wall, and having stones passed from hand to hand along the line. About some ponds there are legends of menehunes* having constructed them.

The ponds are areas of a few to many acres enclosed by walls built across entrances to bays or indentations of the shore where the water is shallow. Or they are enclosed by walls that stretch in graceful arcs between two points of the shore. Some ponds are enclosed pounds where fish were fed and fattened, some have gates to be opened on the incoming, closed on the outgoing tide, made so as to allow small fish to

---

* Menehunes: supernatural people, small in stature, who did prodigious tasks, each task performed in a single night.

pass back and forth all the time but hold back the larger fish. Many ponds have now disappeared through neglect or turning the area to other uses. A few have been filled up by volcanic flows. McAllister (60:28-32) describes the construction of the Oahu ponds. He obtained information concerning 97 on Oahu, the most favoured island in this respect. Cobb (13, pp. 429-430) lists those of all islands in 1900. Beckley (4, pp. 20-21) describes kinds of ponds and how fish were taken from them.

Salt water ponds were used chiefly for storing and fattening the *'ama'ama* (mullet) and the *awa* (milk-fish). Stones with seaweed attached were collected and set within the ponds to increase the food supply of these herbivorous fish (Pukui). Fresh and brackish water ponds were used for the *'o'opu, aholehole,* and for shrimps (*'opae*). In modern times the introduced gold-fish, china-fish, and carp were added (41, p. 375). Already mentioned are the inland streams. Kauai is the island most favoured with streams, and the *'o'opu* of this island were famous. Inland ponds were built along stream beds, so that water could be diverted through them; others were near enough to the shore to allow tide water to seep in and out. The only large natural inland lagoon was Pearl Harbour (old name: *Pu'uloa*) and it was famous for its fish and fishponds (80). Inland pools, which were more numerous before waters were led off for modern irrigation projects, constituted natural ponds. Some of these pools were reserved for the exclusive use of chiefs. One more storage place for fish was the taro patches. Taro was so much used as food that almost every valley floor was devoted to the cultivation of this plant. Wet land taro demanded fields actually under water constantly flowing. By an elaborate system of irrigation, waters were led from the mountains down through these acres of taro fields. The taro patches made excellent subsidiary fish ponds for *'o'opu* and shrimps which could be caught as needed. Mrs. Pukui tells of the exciting times children had when a patch was drained to be replanted. They were allowed to go into the pond to mill about, stirring up the mud as much as possible. This did not suit the *'o'opu,* which demanded clear water for breathing. Up would pop their snouts as the water got roiled and there was a merry time catching the fish as they emerged.

APPORTIONING THE CATCH

The first fish caught was always reserved for the gods and offered on the altar of the fish god on shore as soon as the canoe landed. McAllister (60, pp. 15-16) noted many remains of *ko'a*, or fishermen's temples, along the shores of Oahu; Bennett (7, pp. 48-49) has noted them on Kauai. After fish were offered, or set aside for offering by giving them to the priest, the best fish of the catch were set aside for the chief in an amount to provide generously for his personal needs and those of his numerous household. Then the various *kahuna* (recognized experts in branches of learning), next the *konohiki* (chief's agent and overseer), and finally the people received their share. Division was made according to need, rather than as reward or payment for share in the work of fishing. Thus all were cared for. Anyone assisting in any way had a right to a share. Anyone who came up to the pile of fish and took some, if it were only a child, was not deprived of what he took, even if he had no right to it. It was thought displeasing to the gods to demand the return of fish taken without the right. What Hawaiians thought sometimes about this inevitable sharing of a hard won catch may be known from the following lines from the legend of Niho-o-leki (25, Vol. 1, pp. 492-494) :—

The current is flowing towards Maka'ena,
Where swarm the aku,
Where the giving would be a pleasure,
When the worthless could have a share,
When the hungry of the uplands of Waiahulu could have
a share.

Sharing the catch had one restriction—what was taken was supposed to be for one's own use. Absent members would be cared for. Perhaps a messenger, a child, or relative would ask for the share and take it to the absent member.

FOOD EXCHANGE

Shore localities differed greatly in varieties of fish, and some became famous in chant and tale for their specialties. Variety in foods was appreciated and journeys were made to get that for which there was a wish or craving. Chiefs had only to command, and servants procured what was wanted. Commoners procured what they wanted by a kind of courtesy barter, usually within their own 'ohana, or related community. Handy and Pukui (34) have defined the 'ohana:—

> The fundamental unit in the social organization ... was the dispersed community of 'ohana, or relatives by blood, marriage, and adoption, living some inland and some near the sea but concentrated geographically in, and tied by ancestry, birth and sentiment to a particular locality, which was called the 'aina ... Between households within the 'ohana there is constant sharing and exchange of foods and of utilitarian articles and also of services, not in barter but as voluntary (though decidedly obligatory) giving. Ohana living inland (ko kula ūka), raising taro, bananas, wauke, and olonā, and needing coconuts, gourds, and marine foods, will take a gift to some 'ohana living near the shore (ko kula kai) and in return will receive fish or whatever is needed. The fisherman needing poi or 'awa will take fish, squid or lobster upland to a household known to have taro, and will return with his kalo or pa'i-'ai (hard poi). A woman from seaward, wanting some medicinal plant, or some sugar cane perhaps, growing on the land of a relative living inland will take with her a basket of shellfish or some edible seaweed and will return with her stalks of cane and her medicine. In other words, it is the 'ohana that constitutes the community within which the economic life moves ...
>
> The pivot of the 'ohana is the haku (master, director), the elder male of the senior branch of the whole 'ohana. The haku divided the catch of fish amongst the households of the 'ohana which had participated in the fishing; he presided over family councils; and in general he had authority over the individuals and households in all such matters as entertaining strangers and welcoming the ali'i, in supervising work, worship and planned communal activities. The haku was functioning head of an 'ohana. The term ... has no relation to class, politics or occupation. There were haku of ali'i (chiefly) families, or kahuna families, or fishing and planting families ...

In the absence of the haku, the po'o lawai'a made a division of the catch, according to Mary Pukui. The ohana system was effective in food distribution and is doubtless the cause of a seemingly entire lack of the commercial instinct, and indeed understanding of the principles of commerce in the old Hawaiians. The courtesy barter system has lasted up to the present time as a way of dealing between friends. Yet it is also true that Hawaiians caught the idea of trade very early. Menzies, botanist with Vancouver, who visited Hawaii in 1792-1794, records the following (68b, p. 177) :—

> When the fishing canoes came into the bay in the evening, we had an opportunity of observing their manner of traffic with one another, as the whole village and people even from other

villages flocked about them and a brisk market was kept up till they disposed of all their fish for small nails and bits of iron, and sometimes we observed that they drove very hard bargains. Of these nails the fishermen make their fish hooks, and no doubt are obliged in their turn to purchase potatoes, yams, cloth, etc., from the planters. Thus we find that nails and bits of iron here answer all the purposes of money and circulate amongst the natives in the same way that gold and silver does with us.

A system under which the exchange of goods between commoners depended on courtesy laid itself open to tricksters or practical jokers, as proved by the stories which follow. One story, found in the Fornander collection (25, Vol. 2, p. 426) concerns a man who saw some people coming down from the uplands to exchange foods. He saw them burdened with sugar cane, bananas, " and all else." So he made a feint of having just come in from fishing. He pushed his canoe out into the water—unseen, we presume—then returned to shore and dried his net, hoping he would be seen. Evidently the mountain people were taken in, and brought him their goods, " with the thought that they would receive . . . However, there was nothing received because they discovered that he was not a fisherman, so the barterers lost." Just how they retaliated is not recorded.

The second story is of Kamehameha I. According to this unrecorded tale from Mrs. Pukui, he made an agreement with a man of Kahuku (in the district of Ka'u, island of Hawaii) that " for one calabash of poi, one fish," and the man understood that he would get one calabash of fish. He went to the uplands, filled his calabash with poi, and came to Kamehameha, who gave him not one calabash of fish but one fish, and that a little one. Unabashed, the Kahuku man tied his one small fish to his carrying pole and went off home. All the way along his road people laughed at his one little fish dangling from his carrying pole. Our hero came again to Kahehameha with his calabash, the contents neatly covered with fresh ti leaves. He approached in the humble manner of a subject of the great king, crawling up to his presence, and he set the calabash before him. Kamehameha lifted up the ti leaves and beheld not a calabash of poi but one taro within. (Taro is the vegetable from which poi is made.) The king took the play in good part and laughed loud and long and said that they would have no more one-sided bargains. This little story is the foundation of calling the locality where the man lived Kahuku-kau-'ao-'ao (one-sided Kahuku).

SUPPLY AND CONSERVATION

A knowledge of the abundance of fish and the supply procured for consumption in pre-European days can only be partially arrived at, as well as whether the supply decreased after discovery of the islands and the great changes which occurred in ways of living for the Hawaiians. With most of the population (at least 100,000) devoted to procuring enough food to eat, and considering what hearty appetites outdoor living created, it is likely that the amounts of fish obtained and needed were very large indeed, especially as land animals were not counted on as a steady part of the diet.

It is but natural to look back on the " good old days " and think of them as better. It is what some Hawaiians did as time went on after the irrevocable changes had occurred in their ways of living. In the native newspapers, the question was often asked, " Why are fish so scarce and prices so high?"

How abundant had been the supplies? Some impressions of lavish supplies for feasts are vivid. Certainly in later days there were few feasts as abundant as that recorded in 1814 by Manini,* companion and friend of Kamehameha I— " they caught among them all about 50,000 fish, and between men and women, about 10,000 were present " (71). Kamakau, a Hawaiian historian, states (47, Chap. 4, p. 34-35) that:—

> In old days 400 *aku* (bonito) might be caught with the bait from a single *mālau* (small canoe to hold live bait on deep sea expeditions), and when the double canoes, fleets of single canoes, and large, single canoes came to shore, there was trading, peddling, selling for poi, for pounded taro, awa root, tapa . . . until the fisherman was well provided . . .†

A large haul of a big deep-sea net, such as the *ho'olewalewa*, would fill 10 to 20 canoes.

Kamakau also makes the statement that sometimes with good luck fishermen secured " so many that they rotted and a stench arose and they had to be fed to the dogs and pigs, for

* Don Francisco de Paula y Marin.
† Actual trade commenced after European contact.

there were too many for slicing, salting, and drying. Some
were even used as fuel to cook the others." (*ibid.*, pp. 7-8).
The statement about using fish as fuel is borne out by Mrs.
Pukui and her mother, Pa'ahana Wiggin. In Ka'u, a dry
land where wood is scarce, an overabundance of fish, as when
a large school was caught, was dried and devoted to fuel,
because of its oil content.

Kamakau makes note of especially abundant years,
1830-1831. (48, Chap. 18) :—

> The Lord had blessed the land during those years. Fish
> were so plentiful, especially at Waialua and Waianae, that pigs
> and dogs feasted on those that rotted. On Hawaii and Kauai
> there was the same abundance. The fish caught were the uwiwi,
> the a'ua'u, opelu, akule, alalauwa, kala, welea, kalaliilii(also
> called pahikaua), he'e kukulli . . . At Molokai, kawakawa, aku
> and ahi were simply washed up on the beach and flying fish came
> in huge schools. At Wailua the kahala fishing grounds were so
> rich a man could catch as many as 20 to 40 at one haul. Perhaps
> this blessing upon the land was in compensation for the diffi-
> culties into which the government was falling and the extinction
> of the old families of chiefs and commoners which occurred at
> this time.

It is difficult to guess what the everyday consumption
was. But the impression remained among Hawaiians of
later days that there had been plenty. One Hawaiian writes,
in 1923 (75:43), that: "This (matter of fish supply) is
going to be an important question for several generations, to
understand why there was so much fish in the days of our
ancestors and so little in our time although much meat and
fish is now imported to help supply the people with food."

To conserve the supply of all resources was constantly
in the Hawaiian mind. When plants were taken from the
forest, some were always left to replenish the supply.
Replanting was done without fail at the proper time as beds
of taro and sweet potatoes were used. Fishing grounds were
never depleted, for the fishermen knew that should all the
fish be taken from a special feeding spot (ko'a) other fish
would not move in to replenish the area. When such a spot
was discovered it was as good luck as finding a mine, and fish
were fed sweet potatoes and pumpkins (after their introduction) and other vegetables so that the fish would remain
and increase. When the fish became accustomed to the good
spot, frequented it constantly, and had waxed fat, then the
supply was drawn upon carefully. Not only draining it

completely was avoided, but also taking so many that the rest of the fish would be alarmed. At the base of this action to conserve was the belief that the gods would have been displeased by greediness or waste.

Tabus were an instrument in the conservation programme. The political power was concentrated in the upper class, the chiefs, and the laws of the land and of the sea were their edicts. The penalties for breaking tabus were heavy, often the death penalty for what seems to us a trifling fault. This held the people in a strict discipline. Besides tabus, the relationship with the gods was a powerful determinant of action. The lesser gods that each person had, personal gods, as well as the greater gods whose power was universal, were ever present. Their will was interpreted through the priests, but understood well by the people too. To conserve resources was a custom rigidly adhered to. It was the will of the chiefs, and also the will of the gods, and it was obviously wise. When a man broke this law he expected punishment from the chief's agent (*konohiki*), if his act was detected, but punishment from the gods certainly, for no knowledge was hiddden from their perception. Man appealed to his gods for good luck, but the gods expected man to do his share in making it possible.

Besides the rule of taking only part of a supply of fish, fishing was prohibited during the spawning seasons. Perhaps the most important and well-known tabu of this sort was that governing the *aku* and the '*opelu* (ocean bonito and mackerel), deep sea fish that move in schools and were taken in bountiful quantity when the schools were running. Manby, one of Vancouver's officers, notes (64, p. 8-9) : " The present taboo Bower (?) is an invocation to the god that presides over fish; it is annually observed at this season of the year, as a notion prevails, were this ceremony neglected, the finny tribe would immediately quit the shores of Owyhee." These two fish were important in the fish supply and, as Vancouver observed (85, Vol. III, 18) : " These are not lawful to be taken at the same time." Malo (63, p. 251) states that: " For six months of the year the '*opelu* might be eaten and the *aku* was tabu, and was not to be eaten by chiefs or commoners. Then again, for other six months the *aku* might be eaten, and the '*opelu* in turn was *tabu*. Thus it was every year." The *aku* and '*opelu* were almost sacred fish as, according to tradition, they had saved an early voyager coming to Hawaii from Tahiti from storms at sea by quieting the waters. "When the wind kicked up a sea, the *aku* would frisk and the *'opelu* would assemble together, as a result of which the ocean would entirely calm down." (63, p. 25.) But the motive for placing the tabu was probably to protect the supply during spawning and juvenal season.

Mrs. Pukui gives an interesting account of the tabu system governing fishing seasons in her district, Ka'u, in Hawaii:—

> There was never a time when all fishing was tabu. When inshore fishing was tabu, deep sea fishing (*lawai'a-o-kai-uli*) was permitted, and vice versa. Summer was the time when fish were most abundant and therefore the permitted time for inshore fishing. Salt was gathered at this time, also, and large quantities of fish were dried. Inland crops were tilled, and supplies from the higher lands procured. In winter, deep sea fishing was permitted, and the sweet potatoes that grew in large patches near the shore were cultivated. A tabu for the inshore fishing covered also all the growths in that area, the seaweeds, and shellfish, as well as the fish. When the *kahuna* had examined the inshore area, and noted the condition of the animal and plant growths, and decided that they were ready for use, that is, that the new growth had had a chance to mature and become established, he so reported to the chief of the area, and the chief ended the tabu. For several days it remained the right of the chief to have all the sea foods that were gathered, according to his orders, reserved for his use, and that of his household and retinue. After this, a lesser number of days were the privilege of the *konohiki*. Following this period the area was declared open (*noa*) to the use of all.

There are not sufficient records to tell whether districts varied greatly in this respect, or whether the rhythm of the seasons varied greatly in separate areas. The nature of the fish population doubtless necessitated special tabus in some areas.

For some fish protection was unnecessary. According to one writer (75, 43) :—

> Fish such as the manini, the kole, the uhu, the kumu and the palani and the kala and many others went into sea pools to live until the tiny fish were grown. No kapus were imposed on them at the spawning season. The mullet, squid, aku, opelu and other fish bore their young in a place that was not sheltered ... They were made kapu when the spawning season was near until the months for this duty were over.

In 1900, that is, not long after Hawaii became a territory of the United States, Congress provided for an investigation into " The entire subject of fisheries and the laws relating to the fishing rights in the Territory of Hawaii." (13, pp. 353-499). The published results of this investigation are the most detailed and comprehensive work that has been done on this subject. Jordan and Evermann, who with Cobb, carried on the work, reviewed the old laws (*ibid.*, 359-370), and suggested certain regulations (*ibid.*, 372) which have not been fully carried out to this day.

It is evident that the earliest laws were a carry-over of the tabus. Kamehameha III promulgated the first written set of laws in 1839, and a lengthy section is devoted to fish and fishing grounds. The king relinquished some royal and chiefly rights at that time, and made a division of the fishing ground, (*ibid.*, p. 361) " one portion to the common people, another portion to the landlords, and a portion he reserved to himself. These are the fishing grounds which His Majesty the King takes and gives to the people: the fishing grounds without the coral reef, viz., the Kilohee grounds, the Luhee grounds, the Malolo grounds,* together with the ocean beyond."

Continuing this excerpt from the laws (p. 361) :—

But the fishing grounds from the coral reef to the sea beach are for the landlords and for the tenants of their several lands, but not for others. But if that species of fish which the landlord selects as his own personal portion should go on to the grounds which are given to the common people, then that species of fish, and that only, is tabooed; if the squid, then the squid only; or if some other species of fish, that only and not the squid. And thus it shall be in all places all over the islands; if the squid, that only; and if in some other place it be another fish, then that only and not the squid.

Rules and regulations follow, attempting to protect each class from abusing another. The tax officers for the king were armed with power of placing tabus " at the proper

* The meaning of these terms is: *kilohe'e* grounds—the area shallow enough for wading, or examining the bottom from a canoe, perhaps with the aid of the oiliness of pounded *kukui* nut to smooth the surface of the water; the *lūhe'e* grounds—the area where the water was too deep for the bottom to be in sight and the *he'e* (octopus) had to be caught by line and cowrie shell lure; the *mālolo* grounds were certain rough and choppy areas, crossed by currents, where the *mālolo* (flying-fish) habitually ran. These were deep places, but were not considered the open ocean. (Pukui, inf.)

season " upon certain fishing grounds " which are known to the people to have shoals of fish remaining upon them." A list of such fishing grounds is enumerated for Oahu and Molokai. For the other islands, special fish were subject to tabus, if the amount were over one canoe load.

> In Lanai, the bonito and the parrot-fish. On Maui, the kulekū* (akule-kū) of Honua‘ula and other places. On Hawaii, the albicore. On Kauai, the mullet of Huleia, Anehola, Kahili, and Hanalei, and the squid and fresh-water fish of Mana, the permanent shoal fish of Niihau, and all the transient shoal fish from Hawaii to Niihau . . .

The law ends with this statement (ibid., p. 362) :—

> But no restrictions whatever shall by any means be laid on the sea without the reef (evidently meaning outside the areas already mentioned) even to the deepest ocean, though those particular fish which the general tax officer prohibits, and those of the landlords which swim into those seas, are taboo . . . fine is specified above.

Because chiefs had absolute power, they could relax their own tabus when they wished. Ka‘ahumanu (favourite queen of Kamehameha I) was devoted to the people. Of her, Kamakau (48, Chap. 19) says:—

> " In certain years she allowed the people to fish in the tapu waters of Oahu and forbade the landlords to prevent them from taking fish usually restricted for the chiefs, such as the uhu, opule, he‘e and kahala. For a time there were no tapu fishing grounds for Oahu."

Modification of the laws continued as time went on. The transient shoal fish were defined: akule, ‘anaeholo, alalauwā, uhu-ka‘i, kaweleā, kawakawa, kala-kū. The fish which each landlord set aside for himself had to be proclaimed so that it was clear to the people which fish was tabu, and the director of the government press was required to publish a list of these fish and the localities in which they were tabu. Finally, in 1845, this privilege of having exclusive right to one kind of fish was exchangeable for the right of tabuing all fish over a landlord's fishing ground for a certain length of time. The "royal" fish were defined— certain fish in certain localities noted for their abundance. " These shall be divided equally between the king and the

---

* The adjective suffix kū is added to any fish name to indicate a stand, or pause of the school in its journey. Literally, kū means to stand.

fisherman." (13, p. 364). An interesting point covered was, "On all the prohibited fishing grounds the landlords shall be entitled to one species of fish and those who have walled fish ponds shall be allowed to scoop up small fish to replenish their ponds."

In 1850, it was recognized that "Fish belonging to the government are productive of little revenue," and the laws caused considerable trouble, therefore all the government fishing grounds and fish tabu to the government (the King) were turned over to the people. In 1859, the laws were codified, and still the principle of conservation was adhered to, "for the protection of . . . fishing grounds and minister of the interior may taboo the taking of fish thereon at certain seasons of the year" (13, p. 367-368).

In 1927, Jordan and others (42, p. 650) wrote:—

> The fauna of the reefs is much less abundant than in the period of the first extensive explorations, those of Dr. Oliver P. Jenkins, in 1889, and of Jordan and Evermann in 1901. Probably no species had been exterminated by overfishing, but many once common have now become rare.

Many factors influenced the problem of getting an abundance of fish food for Hawaiians in *haole* (literally foreigners—post discovery) times, the let-down of the strict discipline of the tabu system, with which the replacing *haole* forms of laws and government control did not coincide perfectly, the bringing in of other racial elements, as the Chinese and Japanese, who had strong commercial instincts, even to holding the price by limiting the supply in the markets, a scheme completely foreign to Hawaiian concepts, and the changed condition that took many Hawaiians away from shores and into a more urban way of living.

It is evident that the principle of conservation was a strong factor in Hawaiian sea-food economy. Some of the elements of the old pattern carried over far into European times, but the pattern had to receive constant modification in the hybrid culture. It has not yet been made to fit perfectly.

Certain foods were tabued to women, among them some fish. Cobb was informed (13, p. 360) that these were the *kūmū, moano, ulua* and *hāhālua*. Pogue (73, p. 37) states:—

> It was not proper for the women to eat these: pig, bananas, ulua fish, kumu fish, niuhi shark, whale, nuao fish, hahalua fish,

sting ray, the haelepo, and other things too numerous to be
counted by man. A woman was killed if caught eating these
things.

Malo (63, p. 52) lists: *ulua, kūmū, niūhi, nāiʻa, whale,
nūao, hāhālua, hihimanu,* and *hailepo.* During a woman's
pregnancy, other fish were tabu also—*aku, ʻopelu,* mullet,
or any other white-fleshed fish, or any fish salted by others.
It was forbidden to her to salt fish herself, for if any
mishap occurred and the fish turned out to be improperly
salted so that it did not keep, such inefficiency resulted in a
calamity to the child. It would be apt to have a periodical
catarrhal condition of the nose that was most unpleasant.
It was also forbidden for pregnant women to string fish.

It was not forbidden to women, evidently, to touch
tabued fish. Kamakau says (47, Chap. 4): "The women
separate the tapu food for the men's house from those for the
women's eating house." For men and women to eat together
at any time was forbidden.

When seasonal tabus were lifted, women had to wait
longer than men. Malo (63, p. 196) describes the elaborate
rules that governed the observance of the *makahiki* festival
which Handy (33, p. 296) calls the first fruits and harvest
festival. At the close of the period of days which covered
the ritual, the tapus were gradually lifted. On one day fish
was caught, and the "male chiefs and the men ate of the
fresh fish . . . but not the women." Several days later, "the
queen and all of the women ate of the fresh fish from the
ocean. This observance was called Kala-hua." Even the
female gods had to wait until men were satisfied. As
Kamakau says (47, Chap. 4, p. 62) :—

> (In kala fishing) the head fisherman acted as the kahuna who
> prayed to the aumakua for their blessing. When the prayer had
> freed the kapu the fishermen from one end of the land to the
> other ate; and when they had eaten then they gave fish to the
> women's house for the female aumakua.

Vancouver (85, Vol. III, 18-19) makes note of the tabu of
the aku, "it should cease with the men on the morning of
the 4th, and with the women on the day following."

METHODS OF PREPARING.

Fish were eaten raw, cooked, or salted and dried. Some of the smaller fish, free from hard scales and sharp fins, were eaten just as they came from the water. Great pride was taken in serving raw fish alive, and many tales include an instance of a runner bringing fish a long distance and delivering it alive. One example is given:—

> The chiefess yearned again for the *opule* fish that hides in the sand of Kalapana and the fat *aholehole* fish of Waiakolea (Puna, Hawaii). It is far from Hilo to Puna, but because the chiefess had a craving the distance was as nothing. The *opule* fish was brought to her alive, still breathing, inside of a wrapping of *pakaiēa* (sea weed). The *aholehole* fish was still moving, wrapped in some *limu-kala* found in the pond. (75, 42.)

Manby, one of the crew of Vancouver's vessel, notes (64, p. 9) that Kamehameha ate fish " in the same state as when taken from the water."

Raw fish was and still is considered delicious. Europeans who have learned how to enjoy reef fishing occasionally follow the Hawaiian example of taking a bite from the back of a small fish for refreshment on the reef and have been surprised at how good is the taste. Ellis, Captain Cook's surgeon, notes (21, Vol. II, 167-8) :—

> They have great abundance of fish, which the women are particularly fond of; they eat them raw, guts, scales and all; and use an immoderate quantity of salt with them.

I doubt that women had any better chance to eat raw fish than men, but it evidently seemed so under Ellis' observation. Occasionally a raw fish, still wriggling, has been known to stick in the throat, causing death from the spines piercing the flesh. To this day Hawaiians eat the whole fish, except the bones. Sometimes skin is removed, or scales, if tough or hard, sometimes the gall bladder, sometimes a choice is made of the internal organs, some discarded. But usually the complete fish is consumed. Many Hawaiians would think the fish lacked flavour if the viscera were removed. This fondness for all parts of fish is general to all Polynesia. Anyone who picks at fish served, discarding the dark flesh near the bones, or the skin of any except a very toughskinned fish, or one strongly malodorous, as the *palani*, was pitied as one who does not know how to eat fish—an uncultivated person.

PREPARATION OF RAW FISH

Fish were scaled (*unaunāhi*) by rubbing with a sharp shell, or a knife made of a split section of bamboo (which is very sharp), or any other sharp-edged instrument at hand. Certain small fish could be scaled by putting them into a container with rough pebbles or heavy beach sand and stirring them about until the scales came off, then rinsed.

There are two ways to split open a fish. For drying, it was slit through the head and back (*kaha kulepe*: *kaha*, to slit; *kulepe*, to flip open). For eating, the head was not cut but the body was slit ventrally, from chin to vent (*kaha 'opū*). Such fish as the *'opelu* were slit this way, salted, then folded together again for packing away—folded back to normal position, or from head to tail. To do this the backbone was dislocated in several places (*ho'a nu'u nu'u*). A cut crosswise of the fish was called *poke*. Mullet, *awa* and *'ō'io* were cut this way for salting. The pieces were chunks, not thin steaks. If the fish were a large one cuts or slashes were made into the chunk along the backbone.

Hawaiians took note that fish stomachs differ. Those that are bag-like are called by the general name for stomach —*'opū*. *Akule, aku* and *'opelu* have *'opū*. Stomachs that are hard and ball-like, or gizzard-like, are called *pu'u* (ball, head or knot). Mullet, *nenue*, and *puwalu* are fish that have *pu'u*. (Compare statement by Jordan, 39, Vol. I:32). The *'opū* was often emptied, the entrails set aside, the bag washed, then filled with fish eyes, and baked in the imu. The contents were considered a great delicacy. *Pu'u* and *'opu* were useful as material for fish relish, or *palu* (see p. 27 herein). The contents of the intestinal cavity were usually eaten, except that the gall bladder (*au*) was removed, if bitter, as in the *kala*. In sizeable fishes the intestines were opened and washed. For salting, the guts were removed. In small fishes they were pressed out; in larger fishes, cutting had to be done.

Taste differed as to blood—some liked it, some did not. If it was wanted, that from the gills was squeezed out on to the flesh, so as not to be lost. If not wanted, the blood was rinsed off. A note by Kukea (inf.) is that spearing is not a good way to procure fish for eating raw, as there is loss of blood.

Fish were never brought to the serving place without some preparation. Fresh fish were always salted, if only a little, the salt allowed to permeate the flesh to some extent. Without salt, Hawaiians were at a loss.

Forty-five years ago, great changes in habits of living had already occurred, but Jordan and Evermann (40, p. 357) noted even then that: " As the natives mostly eat some fishes raw, certain species not of special excellence when cooked are very highly valued by them." And again (*ibid.*, p. 373) that this eating of raw fish actually blocked the progress of legislation aimed at protecting the fish :—

> The chief argument used against protective laws is the desire of the Hawaiian people to eat little fishes raw. Of these little fishes thus eaten one or two, called " nehu " never grow large. On the other hand, it may be urged that the nehu is an important food of larger fishes; that the market value of all which are taken is insignificant, and that the young of the mullet and other fishes of real value are taken and eaten with the nehu.

SALTING

Fish were salted* to various degrees. For eating soon, they were lightly salted (*kahunahuna*), and no fish was eaten raw until it had been salted to some degree. When salt had worked in a little, it was at the '*ū'ū* stage. Just before serving, the excess salt was rinsed off (*kaka*). When well salted throughout it was at the *li'u* stage. When stiff and hard (*wikani*) fish were white with salt (*uani'i*). The process of rubbing in salt is *kopī* (or *kapī*).

Large fish eaten raw were prepared by mashing the flesh with the fingers, a process called *lomi* (massage is the nearest meaning), the fish so prepared called *i'a lomi*. This process stopped short of mashing, the object being to soften and allow the salt or other flavour to penetrate. If the flesh were not soft enough for *lomi*, as in the mullet, *uouōa*, *aholehole* and *weke*, it was cut into small slices or chunks, or left whole, and called *i'a nahu pū* (fish to bite into). In

* Salt was gathered along the shore, where it crystallizes naturally from sea water. Care was taken to keep it clean, free of dirt. Supplies of well dried salt were hung up in covered baskets or gourds, and stored in a dry cave, if there were one available. The passage of a creature of any sort over a container necessitated the throwing away of the whole contents, because the spirit of the creature might have had some harmful effect.

modern times raw onion is sometimes added. Fish used for
i'a nahu pū were always good for cooking in ti leaves.

At modern feasts, the menu is hardly complete without
a dish of salmon *lomi*, salmon (*kāmano*) being imported
from the Northwest Coast of America. This importation
started in very early times. Soon after discovery (1778),
Hawaiians began shipping on as crew aboard vessels in the
fur trade of the Northwest Coast. They took a liking to
salmon at once, and brought it back salted. This trade
became a steady one. A keg of salt salmon was always on
hand in well provided households, those who could afford it
getting the choice bellies (*alo piko*). The tail part, not so
easy to *lomi*, was saved for cooking with greens. The
missionary families found in salt salmon a substitute for salt
cod—the New England standby. Salt salmon,.lomied with
raw onion and raw tomato, as a dish for a feast (*'aha'aina;*
modern term is *lu'au*) did not appear until late years
(Webb and Pukui).

Salmon were used in ceremonies soon after introduction. Because the word *mano* (vast, expansive) appears in
the word for salmon (*kāmano*), salmon were usually selected
for the feast after the first work of a novice had been
completed, as the first mat, or the first *kapa* (bark-cloth).
There was a conviction that the novice's knowledge would
thereby become vast. This belief was current up to about
forty years ago. (Mrs. Pukui.)

DRIED FISH

Fish was preserved for use when storms prevented fishing, or tabu seasons forbade it. There were two stages of
drying—partly dried (*i'a maemae*), for keeping a short time
only, and well dried fish (*i'a malo'o*) for keeping a long time.
Kinney uses another term—*i'a kaula'i*. When well dried,
many fish became hard and stiff (*wikani*) and white with
salt (*uani'i*).

Small fish were scaled, cleaned, soft parts removed, but
heads usually left on. They were generously salted (*kopī*),
and after lying in salt an hour or so were laid on clean
stones. In very hot places, such as the beaches of Ka'u,
Hawaii, the fish dried quickly. Kawena Pukui notes:—

> If the flesh was exposed and the stones were too hot, we
> laid down branches or leaves before the fish were scattered over

the hot rocks. If the skin was left on the fish this was not necessary. Very small fish were strung on a bulrush, a banana fibre, a sturdy grass stalk, or a ti leaf midrib. A string of fish was called a *kaili*, or *kali* . After the fish was dried it was packed in calabashes, and the calabashes hung up in nets.

It took only a few hours to salt small or medium sized fish. If caught and salted early in the morning, they might be well dried by noon. Larger fish might take as much as two or three days of hot sunshine. If caught at night, they were allowed to soak in brine until morning, then dried. Large fish were scaled, if necessary—*aku, 'ahi, kawakawa* and some others have no scales—cut into pieces without removing bones, thick pieces gashed and rubbed with salt, and soaked in brine about three days, then dried in the sun. Fish so salted would keep for two or three years in a dry place. They had to be aired and sunned occasionally to prevent being mildewed. If there were signs of the fish not keeping, and the sunning did not cure it, it was cooked in the imu and redried.

Dried fish were eaten without further preparation or were broiled. If only partly dried they would remain a little moist when broiled. If too dry they were broiled, then laid in a dish of water to soften.

Captain Cook observed the Hawaiians' great fondness for salt fish. " Their fish they salt, and preserve in gourd-shells ; not, as we at first imagined, for the purpose of providing against any temporary scarcity, but from the preference they give to salted meats." (15, Vol. III:141.)

Cook was a careful observer, yet this may be too sweeping a statement—that salted fish were preferred to fresh. The Hawaiians were extraordinarily fond of salt, however. A meal without salt was a sad occasion. It was even possible to make a meal of poi—the starchy mainstay in foods—and salt alone. But fish or flesh without any salt was totally unattractive to the palate.

PREPARATION OF COOKED FISH

Certain fish were preferred cooked, such as mullet (*'ama'ama*), *moi, weke,* and *kūmū*. Methods of cooking were by baking, broiling, broiling in leaves, and by putting hot stones in a container with the food. A brief description of these methods follows:—

1. Baking (*kalua*). This was and still is done in an earth oven, or *imu* (old spelling is *umu*). The oven is prepared by digging a hole in the ground sufficiently large to hold the food to be cooked. A fire is carefully laid and on it are placed stones, about the size of a man's fist, to the depth of two or three stones, or enough to fill the hole. Stones must be carefully selected and tested so that they do not burst in a fire (too close grained) or crumble and so lose the heat (too porous). The fire should be so laid as to burn briskly and heat the stones red hot. Embers are then removed, the stones poked about until a fairly smooth surface is made. A thick layer of banana leaves, or a layer of banana trunks, split lengthwise, is laid over the hot stones, then more leaves, banana or *ti* (*Cordyline terminalis*), and then the food to be cooked. All kinds of foods are put into the *imu* together, arranged carefully so that greatest heat reaches food that needs it most. If a fish is to be cooked, it must be wrapped in leaves to protect it from being crushed, and flavour lost. The *imu* is covered with leaves after the food has been placed and then with earth to hold in the heat. About two hours is sufficient time for a small *imu*.

2. Broiling (*ko'ala, kunu, pulehu, palaha, olala*).
   "*Koala* is the placing of fish on good hot coals. Raw fish or dried fish are good cooked this way." (52a, p. 160.) No wrapping in leaves is necessary. In the story of Pamano (75, 46), rock broiling is mentioned—"Why not keep the fish that is caught? When we go ashore we shall make a fire on a flat rock until the rock is heated red. Then we shall broil our fish and when it is done eat it with sweet potatoes."
   Kunu was a term almost synonymous with *ko'ala*, but it implied that great care had been taken in preparation, as it would be for a chief's table. The fish was chosen carefully, was handled so as not to be broken, or burned, the skin not too brittle, the flesh evenly cooked, by turning it often with a stick, or with the hand. It was a source of pride to serve a fish so that it looked almost untouched.
   Pulehu (heaped ashes) was cooking by shoving the food into a heap of hot ashes and embers.
   Palaha (flattened out), a term used chiefly for land animals— broiling a flattened out piece of flesh.
   Olala was broiling by holding over the coals and turning so that all sides were heated. Dried fish did not need actual cooking, merely heating a little. (Note: The Hawaiians of the old days could stand a great deal of heat in handling hot objects. In packing an *imu*, a container of water was handy. With a dip of the hand into the water first, very hot stones could be handled with lightning speed.) (Mrs. Pukui.)

3. Steaming in a closed container (*hakui, puholo*).

These terms seem to be synonymous. The method is to put some small, red hot stones (*eho*), perhaps an inch or so in diameter, in a closed vessel with food. Kepelino describes this method (52a:160) :

" To cook a fresh fish, or any such food, put it into a calabash and put hot stones over it, then lay another fish on top of the stones and close the calabash tight to keep in the steam. After a time, take off the lid, sprinkle on a little salt, pour in a little water and cover again. After a time open and eat. That is the way to cook with hot stones. Such food is savory and delicious. Unsavory is the white man's food "

4. Broiling wrapped food (*lawalu*).

This method was used a great deal. Fish that had been cut into pieces, or small fish that would be lost in an *imu*, or burned crisp if broiled, were wrapped in leaves of the *ti* (*Cordyline terminalis*), occasionally in leaves of the wild ginger, which is said to have added a delicious fragrance to the fish. The leaf bundle was toasted over the open fire, turning it occasionally and the fish was cooked when the juice ceased to drip from the bundle. With what skill this was done has been recorded by Duhaut-Cilly (18, vol. 2:314) :

" These worthy people received us like friends from child-hood and neglected nothing in having made for us a fine supper; and, indeed soon served an excellent repast, chiefly on the delicious mullet which might be called a domestic fish, cooked with such perfection that when the banana leaves in which it had been steamed were taken off, it had received hardly a slight alteration in form and color." (translation)

WRAPPINGS (*laulau*)

The *ti* (*ki* is the old spelling) leaf was a most useful article to the Hawaiians in caring for food. The leaf is long and wide (20 in. x 6 in. is an average size), smooth, shiny, tough, and, except for the midrib, the veins are unobtrusive. It has no odour and is clean and fresh looking. Ingenious ways of folding and tying produced several types of carriers for small foods, and wrappings for foods to be cooked.

Small fish were wrapped in a *laulau pi'ao*, larger fish in a flat bundle called *laulau lawalu*. Fish were always placed on the shiny, upper side of the leaf. The *pi'ao* was made as follows:

Double a leaf over on itself, tip to stem, but twist the leaf so that the shiny side remains uppermost. This makes a cornu-copia, the point of which should be a little to the side of the midrib, on the side toward which you have turned the leaf. Prepare another leaf in the same way. Hold it in the left hand

and in it place the small fish and a little salt. Use the first
leaf as a cover, letting the two telescope. Hold the stems and
ends firmly together. The resulting bundle should be about as
wide as a *ti* leaf. Take a third leaf—more if the stones or embers
are very hot—and fold it over the bundle as reinforcement,
letting the stem and tip end join those of leaves 1 and 2, and
tucking in the points of the cornucopias under this top leaf.
Use another leaf as well if the stones are very hot. When all
is snug, tie stems and ends securely together by binding with
a faded *ti* leaf twisted a little into a sort of rope. (When faded,
ti leaves are supple yet still strong.) Terms for this leaf used
for tying are *hiki'i*, or *nakinaki*, or *hipu'u*.

The method of binding is to form a loop of the *ti* leaf rope,
hold it with the left hand against the spot where the tie is to
be made, and head it toward the pocket end of the bundle. With
the right hand make two or three turns of the rope around the
stems, ends and loop. Pass the working end of the rope through
the loop, then take the other end and pull. This reduces the
loop and makes a very tight knot. Pull as tightly as possible.
In cooking, the leaves will shrink a little and a loosely tied
bundle will be difficult to lift off the hot stones without spilling
the contents.

In olden days families often shared an *imu*. To distinguish
laulaus a particular kind of knot or series of knots was made
in the long end of the binding leaf after the bundle was tied.
Thus disputes were avoided.

Another way of binding was by splitting one of the stem ends
of a leaf of the bundle, and using these ends to twist around the
stems and ends. This was practicable for tying bundles of fresh
foods, but not for foods to be cooked, as the fresh stems were
not supple, and could not be tied tightly.

The flat bundle (*laulau lawalu*) is prepared as follows:
Place a fish on the shiny side of a leaf, length to length.
Do not attempt to wrap a fish that is so long that there is not
enough leaf all around it to fold over for an inch or two on the
sides, and leave at least that much allowance at the ends for
tying. Cover with another leaf, folding it down under the fish.
Hold the leaves together at the ends. (Two people work best at
this task.) Take a third leaf, a little longer, and twist it in a
long diagonal around the bundle. Keep on holding the ends
securely. Repeat the diagonal process with a fourth leaf, start-
ing from the other end. Tie the ends securely, with a faded
leaf. (Figure 1)

A little salt was always added to fish so wrapped. Fish
so large that they had to be cut up were wrapped, as well as
small fish. A few excerpts from tales are here given,
mentioning ways of cooking fish.

He lit a fire for cooking the fish. The *anae, uoa,* and *weke-lao* were broiled and cooked in *ti* leaves. The *palani-maha-oʻo* was left uncooked. (75. 23, March 19, 1914)

He asked his brother to cook the fish, that is the *palani-maha-oʻo* in a calabash. The brother lit the fire quickly, placed some *eho* stones on it, cut up the fish, put it in a calabash, added water, then the hot stones, and put the lid on the calabash. (75. 23, Mar. 1, 1911)

The calabash used for cooking was any wooden vessel, shaped like a bowl with a somewhat narrowed top, and a wooden cover. Hawaiians cultivated many varieties of gourds. If wooden calabashes were not available, these were used instead, but were less durable. Custom forbade that a vessel once used for food be used for any other purpose.

PRESERVED FISH

Fish preserved without cooking made a dish called *palu.* Certain fish were preferred, such as *ʻamaʻama, manini, ʻopelu, akule, kawakawa,* or *aku.* Sometimes the meat from the head only was used, together with the visceral organs, after the gall bladder had been removed. The stomach, intestines, and all other parts were cleaned and added, seasoned with chili peppers,* roasted *kukui* nuts† and sometimes green seaweed (*limu ʻeleʻele*), the whole mixture chopped fine and set away in a closed vessel for several days to ripen. The ripening process was sometimes hastened by letting the fish remain unsalted overnight or all day, then head, tail, bones, and skin removed and all the flesh mixed with chili peppers, roasted *kukui* nuts and salt to taste. This was *iʻa hoʻohauna* (to imbue with a fishy odour), also called *iʻa hoʻomelu*‡ (to allow to begin to spoil) or *hoʻopila-pilau* (to make a stench). *ʻAmaʻama, nenue, manini, hinalea,* and *puwalu* were the favourites chosen for this dish. After setting a few days, *palu* was ready to give piquancy to a meal of greens or *poi.* Octopus, both flesh and ink sac, *aʻama* crab, and lobster (*ula*) were also bases of *palu.*

*Chili peppers were not in Hawaii in ancient times. Don Marin, friend of Kamehameha I, mentions planting them in 1815 (16, p. 193). Hawaiians acquired a taste for them, some even consuming a generous number of them at a meal without giving evidence that they caused a burning sensation. (*Pukui,* inf.)

†*Kukui* nut (*Aleurites moluccana*), a rich, oily nut; roasted it makes an excellent condiment. More than half a nut is a dose that is strongly purgative; when roasted 5 or 6 can usually be eaten.

‡ Some prefer the term *hoʻomelumelu*; meaning identical.

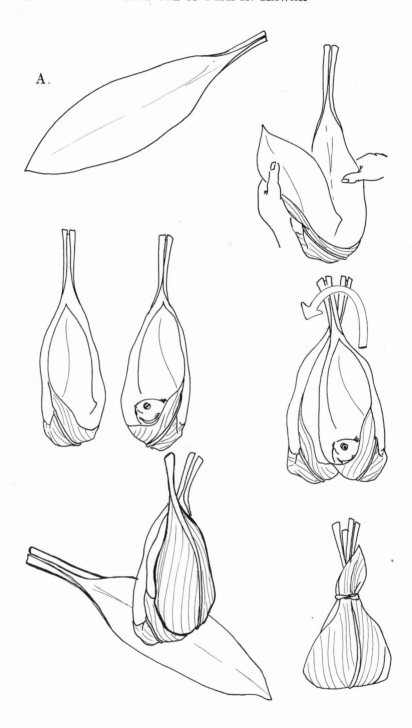

A.

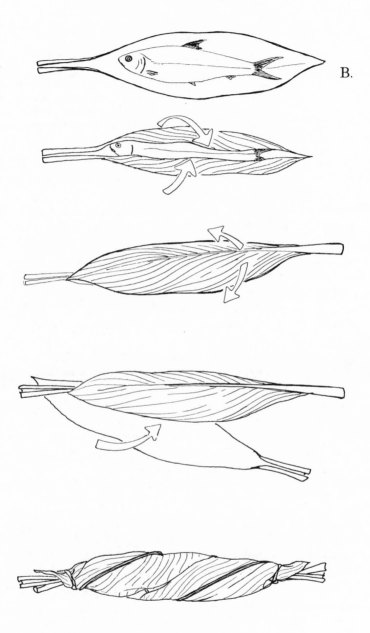

FIGURE 1. Ti leaf wrappings: *A*, Steps in making a *laulau pi'ao* bundle; *B*, steps in making a *laulau lawalu* bundle.

In a contribution on " Some things eaten when the sea
was rough " (75.45) it is noted:

> If the fish was the *'opelu*, the body was eaten, while the heads
> were broken off and stuck into the spaces on the thatching sticks
> of the ceiling. These heads were stuck away to save till there
> was enough, then they were broiled and pounded with *kukui* nut
> kernels that had been cooked. They were pounded together with
> a little chili pepper. When these things were well mixed and
> put into the dish then you might eat it with *poi* and find it
> delicious . . . Our good relishes are vanishing. These relishes
> were well liked by some people, with some they were favourites
> because they were not hard to prepare. One preparation was
> enough for several *poi*-making times.

It is puzzling to know that Hawaiians could eat with
impunity fish foods that had reached a state of deterioration
which is dangerous to Europeans. They relished them, however, and perhaps safety lay in letting deterioration proceed
just so far and then checking it with salt. Mrs. Pukui
suggests that the taste for such relishes may have been
founded on a deep-seated dislike for wasting food, sometimes
difficult to obtain. Occasional experiences of famine made
it clear that eating everything edible was almost imperative.
Habit followed necessity, and taste followed both. However,
this Hawaiian custom may have roots far back in time, before
Hawaiians reached Hawaii and became Hawaiians; possibly
before they reached the east Pacific and became Polynesians.
Should comparative food studies be made for the Pacific and
areas west, it will be interesting to see what connection there
is, if any, between customs of preserving fish, as in the
Indonesian *trassi* and the Filipino *bagoong*.

TASTE IN FISH

All edible fish were eaten and appreciated, some liked
better than others; doubtless people differed in their fondness for fish as much in Hawaii as elsewhere. Fish of strong
flavour were popular, such as the *palani* and the *kala*.
According to Lily Akuna (informant, Puna), the odour
could be dissipated if too strong.

> To remove the odour from such fish as the *palani*, *kala* or
> *puwalu*, which are good to eat but have a rank odour, lay the fish
> across the palms of both hands with the head resting in the left
> hand and the tail in the right. Inhale over the fish from left to
> right, and expel the breath violently. Turn the fish over and
> repeat.

Certain parts of the fish were considered delicacies. Fish were never boned because the flesh around the bones was deemed especially sweet. (Jacobs.) The darker meat next to the spine (*i'o alāea,* or *i'o haku alāea*) of the red-fleshed fishes, such as *'ahi, aku* and *kawakawa,* was especially liked by some. The eyes of large fish were considered delicious, wrapped in *ti* leaves and cooked in the *imu.* " The head is delicious for the eyeballs are there." (75.5.) A liking for fish eyes is general in Polynesia.

Roe was not a delicacy among Hawaiians, and seldom eaten. Because fish were protected by tabus during the spawning season it must have been difficult to get roe, even if the taste for it had been cultivated. The flavour was not generally liked, and when it was eaten it was considered a poor man's dish. One informant speaks up for the roe of the *mahimahi,* however, saying it is delicious when dried and broiled. It was also broiled without drying, and sometimes added to *palu.*

The natural characteristics of each region made some varieties plentiful, others scarce. Hawaiians often became fond of the well-known fish of their region, fads developed, whims of the chiefs varied. Narrators of innumerable tales and meles pause to praise the delicious foods of certain areas. The two excerpts which follow will suffice to indicate that Hawaii had developed epicures and gourmets.

> He, the chief Kakuhihewa, was taken to Ewa, and his personal caretakers were Waipio, Waiawa, and Manana. There were found the fat-bellied *awa* fish of Kuhia, the short-tailed mullet of Olo, the hanging-jawed eels of Hanaloa, the fat *oio* fish of Weloka and the fat mullet of Pauhala. Ewa was the land dearly loved by Kakuhihewa, as was also the warmth of Waianae and Waialua, the tenacious *poi* of Waianae, the delicious *poi* of Kamaile, and the tender mullet of Lualualei. There were also the two pleasant places in Waialua—Ukoa and Lokoea, where there was much fat fish. And the delicious *poi* of Ke-awa-wahie made the chief love these bodies of water. (75.7.)

In the tale of Namakaopao'o (75.40) a *kahuna* is trying to help a man procure a husband for one of his daughters. A likely candidate is now visiting them. The best of everything is desired for the feast, and fish are important to the meal.

> Here is a brave and mighty warrior who is also a divine chief. Kill your pigs without regret, for pigs can be gotten again. Find some well fed dogs with curled tails, and so fat

that water can remain on the back . . . From your ponds the
'ama'ama and *awa* fish as fat as hogs must be fetched. The *'awa*
root growing here too . . . the *hiwa* coconuts, a black fowl, a
white cock . . . and some red fish. Send men to the mountains
for the things that they must bring to you, such as *palai* fern, *ti*
leaves, *ieie* vine, gingers, *kupukupu* ferns, *halapepe, lama, lehua,
kukui* leaves, *ohia* and *koa* . . . Send for sea food, such as red
fish and *aea*, and while some seek these let others gather *limu*
on the shore, some *opihi, wana*, red shrimps, and even if only
four of each are found, that is enough for our purpose. Just
as long as there are enough for this boy's mouth and for yours . . .
After all has been assembled, send your men to Kamo'oloa to
catch *'o'opu* fish, the best liked fish of our land (Kauai). In
these months, Nana, Welo, Ikiiki, the *'o'opu* are fat . . . There
are also shrimps that they catch in basket traps. This is the
last thing to be done—send a swift runner to Wainiha for the
'o'opu nopili that feed on *lehua* petals. There they lie full of
fat and fragrance after feeding on *lehua*. Send your runner this
very day and have them in the house. Do not wait until
tomorrow, for the *'o'opu* is not delicious when too fresh. They
must be caught today, salted and wrapped in ginger leaves,
and the petals of the pandanus blossoms, so that when they are
cooked in the leaves they will be fragrant and delightful to the
nose of the guest.

POISONOUS FISH

In tropical waters some fish are poisonous to eat at
certain times of the year. Throughout many years—the
fact was known long ago—various comments, observations,
and studies have been made of this condition, drawn from
the Mediterranean, the West Indies, and many parts of the
Pacific. It seems that the fish themselves look normally
healthy and show no signs of being poisonous; the only test
so far discovered that detects the poison is " trying it on the
dog," or a chicken, or one member of the group. One or
more fish in a catch may be imbued with poison, the rest
free of it; the internal organs, especially those of reproduc-
tion, carry the greatest poison concentration, and at spawning season the fish are most sure to be poisonous, at other
times they may be entirely safe to eat. Immature fish are
never poisonous. The poison is of the alkaloid type and
affects the nerves, especially those of the limbs. The poison
may be quick and deadly or wear itself off slowly in a matter
of weeks or months. Fish from one part of a lagoon may be
poisonous, those from other areas of the same lagoon safe
to eat. Most writers suggest the poison is derived from

plant or animal growths on which the fish feed. In spite of their keen observation, natives do not know how to detect poisonous fish, and are frequently victims. Not everyone is affected the same way and those of strong constitution are as vulnerable as the weak. The only recorded method natives have found to clear the poison from fish under suspicion is to keep them impounded in sea pools for several weeks. Estuary and deep sea fish are immune, it is the reef fishes that are dangerous. The poisonous quality is present in certain fish only, as the puffers, sea bass, snappers, trigger fish, bonito, mackerel, anchovies, barracuda, possibly others, and each of these fish, if poisonous, presents peculiarities of effect on the consumer.

These are the conditions in tropical waters. It may be that in the shallow seas of Indonesia the condition exists to a less degree, as less evidence has come from that area. Data for these statements is derived from the writings of Anderson (1), Dumeril (19), Grimble (30, pp. 17-19), Gudger (31), Jordan (38), Lesson (58, Vol. 1, p. 225), Maxwell (68, pp. 219-226), Seale (77), Yudkin (90), and Steinbach (78). The full bibliography on this subject is fairly extensive.

Hawaii escapes most of this misfortune, being in the semi-tropics. The poisonous organisms evidently are here on our reefs, but conditions are not favourable to luxuriant growth. The waters about Lanai and Molokai are those most affected, if evidence concerning the *weke* fish is conclusive. Detailed notes as to poisonous quality of a few of the Hawaiian fishes are included in the descriptive list of fishes (pp. 54-139 herein), under the fish affected, namely, *weke*, *uouōa*, *ulua*, *aku*, *'opelu*, *humuhumu mane'one'o*, *'o'opuhue*, *kukala* (called also *hoana*), and *loulu*.

Cases of poisonous fish sickness in Hawaii during the war (1945) came under the care of Doctors Lee and Pang, and have been recorded by them (57). The fish were not from Hawaiian waters, but from islands to the south-west.

The most carefully detailed report made of poison fish sickness from Hawaiian fish (Tetrodon) is that of Larsen (56).

In the minds or experience of some fishermen the effect of moonlight on strongly phosphorescent fish is harmful, causing an itching sensation and fever. These are poison fish symptoms. However, Mrs. Pukui suggests that because

moonlight nights are accompanied by heavy dew fish dried
then do not dry well, and the poison may be bacterial rather
than toxic. Doubt remains because bacterial poison does not
cause an itching sensation.

One way of catching fish in pools was by stuffing the
pounded leaves of certain shrubs under overhanging rocks,
or putting them under stones in the large sea pools (*kaheka*)
or in streams. Fish became stupefied very quickly, and were
easily caught as they floated to the surface. The fish were
cleaned at once, and the stomach and intestines thrown
away. Except for careless delay, there was no danger of
poisoning the consumer. A description of this custom has
been recorded by Stokes (79).

FISH IN BELIEF AND RELIGION
GODS; AKUA

The four most powerful gods of the Hawaiians were Ku,
Kane, Lono, and Kanaloa (5, p. 3). Below them in power
were many other gods, some of them stemming from, or
being divisions of power under the great gods. The god of
fishes and fishermen was Ku'ula (red Ku—red was the most
sacred colour), and his mate was Hina-hele (travelling
Hina). Hina was a female goddess of many names, signifying her duties and powers. Sometimes Ku'ula's wife is
called by another Hina name.

There are several versions of the Ku'ula story, telling
how he gained his place as a god. Fornander collected one
(25, Vol. III:172-174), and Moses (Moke) Manu told the
Maui version to Mrs. Nakuina, who translated it (65,
pp. 114-124). It is here abstracted:—

> Ku'ula possessed a human form and had "miraculous power
> (*mana kupua*) for directing, controlling, or influencing all fish
> of the sea, at will." He married Hina-pu-ku-i'a, and had a son,
> Aiai. After catching and killing a famous eel, an *'aumakua*
> (personal god) that had come from its home in Molokai to
> Ku'ula's neighbourhood in Maui, and played havoc with the fish
> there, he submitted to being burned to death by the chief of the
> land where he lived and had helped the people procure fish
> plentifully, because the chief had been tricked into giving this
> order by the cleverness of a devotee of the eel. All fish vanished
> from the waters nearby as a consequence of Ku'ula's death.
> Ku'ula and Hina were translated into the god state, but, fore-
> seeing all, Ku'ula had given instructions to his son to carry out
> after Ku'ula's death. Aiai escaped, was given refuge by some

kindly people, and repaid the hospitality by bringing back the fish to the district that had been visited with the vengeance of the gods for the death of Ku'ula. Aiai consecrated the first *ko'a ku'ula* (small fishermen's shrines along the shore, where the first fish of a catch is offered), and thus started the practice of making offerings of fish to the god. He went about the islands spreading this recognition of his father, Ku'ula, as the fish god, for it was through Ku'ula's power, with Aiai as his agent, that the fish were persuaded to return.

A quaint story of Kaneaukai (one of the Kane gods) pictures a god as gathering in worshippers as a politician does votes. In searching for his brother and sisters, Kaneaukai came upon two old men who were setting out for the beach to do some fishing. Kaneaukai called out to them:—

> " Say, you old men, which god do you worship and keep?" The old men replied: " We are worshipping a god, but we do not know his name." Kaneaukai then said: " You will now hear and know his name. When you let down the net again, call out, ' Here is the food and fish, Kaneaukai,' that is the name of the god." The old men assented to this, saying, " Yes, this is the first time that we have learned his name." Because of this fact, Kaneaukai is the fish god worshipped by many to this day, for Kaneaukai became their fish god, and from them others, if they so desired. (25, Vol. II:272.)

Along with the worship of Ku'ula was included worship of his family—Hina-hele and Aiai, their son. Vaguely in this set of fishing gods, though not within their family relationship, was Hina-'opū-hala-ko'a (Hina of the coral stomach, meaning Hina of the coral reefs).

GODS; 'AUMAKUA

Ku'ula was the great god of fishermen, but all who fished also depended for luck on their own *'aumakua* (personal gods), and were careful to give thank offerings to them for success.

According to Mrs. Pukui, *'aumakua* were spirits but they usually chose a particular plant or animal as their physical embodiment. The physical form then became tabu to the worshipper, because the *'aumakua* might be within. If an abnormally coloured fish, as an albino, or one abnormally shaped, was discovered it was treated as though it were an *'aumakua,* for it might be one. A fisherman finding such a fish in his net would consider that the

'aumakua had blundered, and he would toss it back into the
water, with an exclamation, "Tsa! What are you doing in
my net!" The "queen" nenue (see p. 97) was probably
considered an 'aumakua. Dr. David Bonnet (pers. comm.)
reports that:—

> When I was in Waialua in 1942, I heard a story of the
> fish pond near Haleiwa. I noticed some albino mullet in the pond
> and mentioned it to a Hawaiian. She told me that this pond
> was the place of the *Mullet God* and that the white fish were
> never removed. When caught in a net they were always released.
> I verified this by talking to the pond operator who told me that
> the white fish led the others into the traps and that he never
> removed them but always threw them back.

The worship of 'aumakua was directed to the spirit, and
along with the worship appeal for magical or godly assist-
ance at whatever was the undertaking. The most common
'aumakua forms were the shark, eel, porpoise and lizard.
Worship consisted of daily prayers and offerings of food.
Neglect brought dire punishment, favour was withdrawn—
a serious matter to a believer. Some 'aumakua dwelt at
prominent points of land along the shore, or inland. If a
species of shark were 'aumakua, offerings had to be made
to any of that species, but many Hawaiians fed a shark, for
instance, at a special spot along the shore and came to know
certain individuals there, and felt sure that the individual
seen was the body that his own 'aumakua had chosen to
enter.

Among well known 'aumakua were Malei, an *uhu*
'aumakua, of the Makapu'u region of Oahu, Kini, a fish god
of Waimanalo, and Kane-kokala. Names of 'aumakua were
no revelation of what animal or plant form they had chosen;
they were mere names. (Pukui.) These gods are described
extensively by Emerson (22, pp. 5-24) and by Beckwith (5).

In gathering notes about the preparation of shark for
eating, Mrs. Pukui had to make inquiries outside her own
family, the shark being 'aumakua to her. Only two kinds
of shark were considered edible by Hawaiians, reef shark
and hammerhead. Mrs. Pukui found a man who knew how
to prepare hammerhead shark—the only kind he ate. (See
under *manō*, p. 93.)

In the story of Paao (pp. 99-100 herein) is given the
legendary reason for the *aku* and *'opelu* being 'aumakua

to all descendants of Paao. A few allusions to fish *'aumakua*
from Hawaiian writings are here given.

In the tale of Laukiekie (75.27) :—

> She lighted the way in the sea of Hina-lau-limukala (Hina
> of the *kala* seaweed) and her husband, Kane-piliko'a (Kane of the
> coral beds). These two were the ancestral *'aumakua* of the
> appropriate sea food used in the medical practices of the kahunas
> of old. Traces of this practice are still here among us to this
> very day . . . These *'aumakua* were called by the fishermen
> " Ka-hale-o-ka-i'a " (the house of fish).

The protective power of *'aumakua* is mentioned in the
tale of Makalei (75.31) :—

> He tossed an *ahi* to the men who had furnished him with
> bait . . . " You gave first, and I caught the *ahi*. My turn comes
> next, to give to you two, and in this way the *'aumakua* of fishing
> will take care of us, their devotees, so that we will always have
> luck."

Emerson (22, pp. 8-12) gives interesting accounts of
various kinds of *'aumakua,* and says :—

> The shark was perhaps the most universally worshipped of
> all the *'aumakuas,* and, strange to say, was regarded as peculiarly
> the friend and protector of all his faithful worshippers . . .
> Each several locality along the coast of the islands had its
> special patron shark, whose name, history, place of abode, and
> appearance, were well known to all frequenters of that coast.
> Each of the sharks, too, had its *kahu* (keeper), who was
> responsible for its care and worship. The relation between a
> shark-god and its *kahu* was often times of the most intimate
> and confidential nature. The shark enjoyed the caresses of its
> *kahu* as it came from time to time to receive a pig, a fowl, a
> piece of *'awa*, a malo, or some other substantial token of its
> *kahu's* devotion. And in turn it was always ready to assist the
> *kahu*, guarding him from any danger that threatened him.
> Should the *kahu* be upset in a canoe and be in serious peril,
> the faithful shark would appear just in time to take him on his
> friendly back in safety to the nearest shore . . . This story of
> shark intervention and many similar to it are extensively believed
> at the present day. (1892.)

FISHING SHRINES (KO'A)

Each shrine was dedicated to a particular god, usually
Ku'ula, but frequently two shrines were adjacent, signifying
one for the male, another for the female god—Ku'ula and
Hinahele. (Mrs. Pukui.)   Many remains of *ko'a* are still

known by Hawaiians. As late as 1930, McAllister (60, pp. 15-16) found that:—

> Although the *heiaus* (temples) have been deserted for a century and even the beating of the drums on the nights of Kane is seldom heard, when fishing is consistently bad in a district the Hawaiians are still known to gather and perform ceremonies and leave offerings at the *koʻa*. The fishing shrines are located near the water, and all of them are small structures. The first fish of a catch was placed on the *koʻa* as an offering . . . Apparently there was one individual who made the offering and who was looked upon as the guardian of the *koʻa*.

" Several *koʻa* were sacred to certain fish," continues McAllister. One site (site 46, p. 68) was " merely a stone at the edge of the water, but it had a great attraction for mullet." This was known as Palielaea, and was at the Waikiki side of Diamond Head. Two others were at Waikiki itself, one named Huanui, for mullet (site 47, p. 68), another, named Hina, for *akule* (site 48, p. 68). Another shrine for *kala* and *nenue* was at " the land known as Kalanai . . . Laʻie." (site 274, p. 156). Another mullet shrine (site 285, p. 158, known as Kauhukuuna, was on the " Hauʻula side of Laniloa Point. A few stones on the beach are all that remain . . ." One more, for *akule,* was found by McAllister on the west side of Kahana Bay. It was known as Kapaelele (site 298, pp. 163-64). *Koʻa* are such simple shrines that doubtless they were thoughtlessly disturbed and the remembrance of many was lost long before 1930. Bennett (7, pp. 48-49) notes:—

> The fishing shrines of Kauai are not notable. They are relatively scarce for the places where they probably once stood are now occupied by plantations and settlements. The principal and simplest shrine is the flat rock on which fish are laid. There is some evidence that this practice still continues . . . Some fishing shrines were probably small houses . . . The chief distinction of a fishing shrine is it position: a rocky point at the end of a ridge, or the terminus of a sand beach.

Considerable fishing was done in the deep waters off the little island of Kahoʻolawe. The archaeology of that island has been studied by McAllister. Nine fishing shrines were found (59a, pp. 13-17). One shrine, at Kamohio Bay, bore signs of considerable age and the remains of many offerings. On one terrace (there were five terraces) was a wooden image within an enclosure consisting of a line of stones. There were also " smooth, water-worn stones in an erect position, some wrapped with *tapa*. South-east of the wooden image there were at least four *tapa*-wrapped stones and two without *tapa*. North-west and back of the wooden image was a ' stick fence ' which was made of sticks wrapped in *tapa*. Behind this ' fence ' and against the cliff were three smooth erect stones . . . A great accumulation of offerings and debris indicates that the site was used for a long period of time . . . Stokes (whose earlier investigations were incorporated into McAllister's work) thinks that the shelter was the abode or workshop of many successive Kahuna Kamakau or fishhook-makers. Every craft had its guardian deity to which of course oblations were made. As time progressed the reputation of the establishment's products spread to the other islands, until fishermen from the islands of Maui and Hawaii resorted to the spot, making offerings to the fish god and bartering for hooks."

This speculation is founded on the great number of half complete or broken fishhooks and squid lures at the Kamohio site.

The archaeology of Lanai has been studied by Emory (22b). He quotes the early missionary, William Ellis, as saying that the small island of Lanai once had a population of about 2,000 Hawaiians (p. 7). Emory found that the *ko'a* were " numerous and varied "; some are rectangular enclosures or platforms, others are circular, and still others are simple heaps of stone. (p. 70.) He describes twelve observable in 1912. Choosing one of his descriptions, we have this picture (p. 71) :—

> On the edge of a cliff at the first indentation of the coast south of Kaumalapa'u is an enclosure 20 by 25 feet, with vertical walls from 4 to 6 feet high, 2 to 3 feet wide. Its walls are of uniform height even on slopes. Its floor is paved with flat stones, and littered with pieces of coral, shells, fish-bones, and charcoal. Joining the enclosure on the north is a house platform 26 by 35 feet, with a sheltering wall 6 feet high on its eastern side; a square stone fireplace is sunk in the middle of this platform, another in the ground 8 feet south of the platform. The presence of charcoal at this *ko'a* suggests recent use.

The archaeology of Maui and Hawaii and Molokai have not yet been completed, but there is no reason not to expect that the shores were dotted with shrines of fishermen. One area of Maui visited by Emory and Maunupau in 1922 yielded evidence of use as late as 1916. Maunupau writes:—

> We went to see the fish heiau . . . (at Nu'u, Kaupo) 26 feet long, 15 feet wide, the stone wall four feet thick and one and one-half feet high from base to top. In the centre is the shrine, 5 feet long and 4 feet wide. There were the stones, the imu and the ashes. There were no coral rocks in this enclosure or on the stone wall, but the walls were covered with fine coral pebbles from the shore.
>
> A hundred feet from here lies a heiau on an elevation on the west side of the canoe landing of Nu'u, close to a rocky base. We found in a rocky corner on the upper side of the heiau an ala stone wrapped in a newspaper, the " Hoku o Hawaii," of September 30, 1916.

McAllister has made the fullest record of the offering bundles which he had the good luck to find at Kaho'olawe, at the shrine of Kamohio or already brought to Bishop Museum. He describes the contents of forty-seven bundles. No two bundles seemed to contain the same objects. (p. 38.) He says:—

> Some contained plants primarily, and little other material. Others were primarily of tapa. Some contained one or two central objects around which other material was wrapped. Others comprised several objects which appeared to be of equal import- ance, wrapped in tapa or grasses . . . Plants of some sort, if only a few leaves or a piece of fern, were used most generally . . . Tapa was found in slightly more than half . . . of the bundles. Bones of birds, animals, and fish were in 49 percent of the bundles . . . apparently no human bones . . .

IMAGES OF FISH GODS

Besides shrines there were sometimes images, either at the shrines, as related by McAllister above, or separate, as well as stones that fishermen relied on to bring them luck. Whether these luck stones were always material dwelling places or locales of the fisherman's 'aumakua, or sometimes that and sometimes merely lucky stones, charms, we do not know. But it is likely that the stones usually stood for the god of the fisherman or an 'aumakua. Sometimes they were carved with human delineaments, sometimes not. Sometimes they were brought along when fishing, sometimes left at home. When left at home the " image " was always placed facing the sea. (Mrs. Pukui.) In the Fornander collection of Hawaiian writings (25, Vol. 3:174) there is the statement, "The fisherfolks' deities throughout the islands were simply certain designated stones; in no case were they carved images." However, at Bishop Museum there are a few stone images carved in the form of fishes, and said to be fish gods. Three of them are grooved as if to provide for the lashing of a cord. They are too small for anchors for canoes, possibly they were fastened to fishlines in place of sinkers, with the idea that if lowered into the sea the god would be in close proximity to the fish he was implored to affect. McAllister speaks of (59a, p. 35), " A number of water-worn elongated stones brought from the shrine . . . undoubtedly *akua* (god) stones . . . all probably stood erect and were embedded in the earth. With several of them fragments of *tapa* or leaves or ferns were associated, undoubtedly all of these stones were once similarly adorned . . ."

Besides formless images or those in fishlike form there are a few with human characters. McAllister found (*ibid.*, p. 36) " five echinoderm spines from the Komohio shrine have been carved—one with unusual skill. The image carved on this spine is typical old Hawaiian and resembles the wooden images that were once placed on the temple walls . . ." The Kamohio shrine also yielded, " One wooden image, now lost, and the head of another . . . The image head that remains was wrapped in many folds of *tapa* and is so badly decayed that the features, a broad forehead, pointed chin, nose apparently long and narrow, pouches under the eyes, and evidence of a protruding tongue, are scarcely visible . . ." This head measures 5.7 inches from the chin to the top of the head. In his notes, Stokes speaks of this image (79a) :—

> " Then we went to what had apparently been a fish heiau of great importance. It was originally a large rock shelter in the front of which, to the north of the middle, had been set up a wooden idol and terraces of stone built up to it." (Ms. notes on Kaho'olawe in Bishop Museum by J. F. G. Stokes.)

Mr. Stokes thinks this image may have represented the god of fish-hook-makers, not fishermen. (pers. comm.)

In Bishop Museum there is one wooden image marked " akua lawai'a " (fisherman's god) from Kailua, Hawaii, as well as the wooden god described by McAllister from Kaho'olawe of very vague features.

Emory found no image on Lanai, but there was the memory of one (p. 71), " upon a great platform near the sea.

A step, one foot high, runs along the front of the altar where once stood an image of the god Kuula, patron of fishermen. Natives who claim to have seen this stone idol, called Kunihi, describe it as two feet high, with ears, nose, mouth and arms. . .   Ohua was one of several men instructed to hide the image by Kamehameha V during his visit in 1868."

Maunupau, who accompanied Kenneth P. Emory on a trip to Kaupo, Maui, in 1922, writes of their finding more than one fishing *heiau* there, (67a) :

> We asked (Hawaiians living there) if they knew what was done at the fish heiau . . . Kenui . . . told us, " The fish heiau was for the purpose of making fish of all kinds multiply, according to one's desire. It was also called ku'ula for a certain god of fishermen and also a ko'a because fine coral was used to cover the spot for the offerings, coral from the beach . . ."

Dr. Emory (pers. comm.) supplements this statement, saying that the natives told them that there was a little house on the *heiau*, that is, a small shelter just large enough to cover a man. A fisherman would sleep in this small shelter and in his dreams would come directions as to where to seek fish.

Another point made was that several fisherman combined in worshipping at a *heiau*.

Emory says that Maunupau's father, a renowned fisherman of Kona, Hawaii, used to offer fish at a *ko'a* upon coming in from fishing—perhaps only 30 years ago.

CEREMONIES

The *'opelu* and the *aku* were two fishes that were depended upon for food to such an extent that they were almost sacred fishes. Perhaps they were so regarded. Both were protected by tabu during spawning season, the open season for one covering the tabu season for the other. In describing the great religious festival or series of ceremonies that occurred at the beginning of each year, the *makahiki*, Malo (63, p.199) says:

> It was in this same tabu-period that Kahoalii plucked out and ate an eye from the fish aku, together with an eye from the body of the man who had been sacrificed. After this the tabu was removed from the aku and it might be eaten; then the opelu in turn became tabu, and could be eaten only on pain of death.

At Kalae in Ka'u, there was a special *heiau* called
Kalalea for the *'opelu*. I find no record of a special *heiau* for
the *aku*, but there may have been one—or more than one.
The lengthy ceremonies which were performed at the time
of the beginning of the *'opelu* season are described in
material found and translated by Fornander (25, vol.III:30-
34). A summary is given here

Offerings were made by the priest to his ancestral god,
and his deity. On one night " fires were restricted, noises
prohibited, also the crowing of the rooster, the grunting of
the pigs, the barking of the dogs. It was a most sacred
night." Prayers were offered by the priest in an inner
shrine, then the people outside, who seemed to have a real
share in this ceremony, were asked to give praise that the
prayers were in good form. Offerings of food—pig, coconuts, bananas—were made, the people sharing, though not
at the *heiau*. Omens were sought by the priests, and prayers
were numerous. Finally,—

> The *'opelu* fisherman then prepared his canoe and his net, and
> at early dawn repaired to the temple to bring the bunch of *pala*
> ferns which he placed in the canoe, at the same time asking for
> a blessing from the deity. He came away after placing the *pala*
> ferns. And when the fisherman saw the priest come away he
> collected together his fishing apparatus and his net. He girded
> on a white *malo*, and chanted to his ancestral deities, saying
> " O ancestral gods of the night, the night is over and I am come
> with the day. Here is the *malo* . . . watch over me that I may not
> be shamed." After this prayer he put his net on board the canoe
> and sailed out to sea. This was a most sacred day, no fires being
> lighted here or there, no other canoes being seen on the ocean
> this day, lest they perish.

The fisherman reached the fishing ground, and praying
all through his labours, cast his net, threw out the bait and
drew in his catch, giving praise to his god, Ku, as he did so.
He and his assistants then proceeded to shore, " making
cheerful noises."

> When they landed the priest came and stood at the landing
> place. The fisherman took up seven *'opelu*, walked up to the
> priest and placed them in the priest's hand. The priest then took
> the *'opelu* to an *uhe* board (offering place on the *heiau*), where
> the fish was consecrated to the deities. The fisherman then went
> to his house to thank his ancestral deities, while the priest took
> the fishes and placed them on a tray before the altar. Then the
> high priest ordered a man: " You take some *'opelu* for the king,
> that he may eat of the first haul of the day." The man went as

ordered, singing as he went. And when the people saw and heard him they all sat down. And when he came before the king he handed him the fishes and then ran away with great speed, lest he be killed. The king then went to the shrine where a priest prayed. They then prepared the king's fish from which the king picked out the right eye and ate it, and offered thanksgiving to the deity.

This does not end the ceremony. Ceremonial fishing is done again, followed by praise for the fishing gods. Another sacred night follows. " In the morning the women were released from their restrictions and were allowed to eat fish. Then canoes came in from the ocean . . . nine days the canoes could not go out . . ." After this we suppose the season was fully open.

Handy suggests (33,p. 229) that the eating of the eye of the 'opelu fish by the king signified giving the fertilizing gods their share and by so doing gaining their blessing on the rest of the season's bounty. It amounted to a thank offering to the god, through his representative, the king. The act of eating established a direct contact with the god, through the king and the object to be obtained—the fish itself.

Whether this important ceremony for the shifting of the tabu from the aku to the 'opelu was carried out at one of the large heiaus inland, or at a fishermen's heiau (having the special name of ko'a) along the shore, is not stated. Little is known of the worship practices at the numerous ko'a. Kamakau described the moment of coming ashore from a fishing expedition (47, Chap. 4, p. 35) :

> As soon as the fishing fleet reached the shore, the head fisherman stepped ashore holding an aku fish in each hand and went to the heiau of Ku'ula where he offered prayer; and when he had finished this worship of the god he threw down the fishes for the male aumakua on one side and those for the female on the other.

FISH AS OFFERINGS TO THE GODS

The best foods, those most rare, and highly relished, were fit offerings for the gods, and rigid rules and regulations determined the choice. This knowledge was vested in the kahuna (priests). Handy suggests (33, p. 329) that

> As regards human sacrifice, the fact that human victims in Polynesia were referred to as " fish ", and were hung in trees

like offerings of fish, has led me to believe that the ancient
form of flesh offering was fish, and that the presentation of
human victims was a late innovation.

Red or white fish seemed to be the kinds most often used.
The red fish were usually the *weke-'ula, moano,* or *kūmū,* the
last being the favourite. "The natives were accustomed to
its being a lifesaver" (75.1, 9/28/1895). It was used as an
offering for any undertaking such as the building or launch-
ing of a canoe, or dedication ceremony in the *hula* school, or
in a ceremony for atonement for sin. White fish used were
the *ahole, 'ama'ama,* or the light-coloured *weke* with a yellow
stripe on its side.

One of the most important offerings was the pig. Sometimes this was unobtainable and then a sea "pig" might be
substituted. Hawaiians had a feeling that each creature of
the sea had its counterpart, or analogous form, in some
living thing of the land—plant or animal. The *aholehole,
awa, kūmū, 'ama'ama* and *humuhumunukunukuapua'a* were
sea pigs (*pua'a kai*). This arrangement must have helped
out considerably when there was no precious pig to be had
for an offering.

Offerings were often chosen because of the magic
inherent in the meaning of names, many words in Hawaiian,
as in other languages, having more than one meaning.
Mrs. Pukui gives an instance:—

> When Namaka (a young relative) and I were initiated into
> the hula school, Namaka gave an offering of *'ama'ama,* while I
> gave a *kūmū,* as I desired to learn not only the dances but the
> background and lore, and thus become a *kumu,* or master.

The following quotations give a few examples of the
purpose and method in making offerings of fish.

To ward off an evil spirit. (75.48):

> They sent a man to fetch these things:—a dried *'awa* plant,
> a whole one of the *hiwa* variety, stalk, leaves, and all. A man
> was sent to the beach for the following,—an *ohiki* crab, an
> *aumoana* crab, and *a'ama* crab, and an *ahole* fish. The prophets
> told the men to bring these sea foods to the sand where the
> canoes of Keliihelehoakahi had departed for Molokai. When the
> *'awa* was brought, the prophets ordered it taken to the shore
> with all the other things necessary for the practice of their
> craft. (Note: Kawena Pukui says, "The reason that these sea
> creatures were chosen was because of the meanings of their
> names,—*ohiki*—to pry loose: *a'ama*—to loosen one's grip; *ahole*
> contains *hole*—to strip away; thus, the evil influence is pried off
> and stripped away.)

To ward off evil from a new house. (75.2) :

This was something done in housebuilding. Some fish were put where the posts were to be placed, that is, the posts on the side facing the east. The fish to use were the *aholehole*, and the *kole*. Should a *kahuna* enter and predict trouble for the household he would die.

To cure infatuation. (75.25) :

This is the way the *kahuna* answered her question. " Thus must you and your parents do to rid you of this sickness. Tomorrow, go to the sea of Halape while it is yet dark and dusky. Go to the beach of Laeapuki and watch for the appearance of the sun—one ray, two rays, and then the third ray. There is a small hollow in the stone, with sea water in it, and a ledge running inward. Let your right hand slip into the water, turn it to one side, grasp the first fish you touch, and hold it. That fish is slippery, slimy. It slips out of your grasp, so I advise you to hold tight . . . Watch the sun, and when it is halfway out of the sea, half still in the sea, raise the fish up close to your mouth and call out in this manner :—

" O Hina, producer of fish,
O Hina, producer of food,
Give prosperity.
Produce a husband who is a ruling chief,
Produce a husband who is ruler over an *ahupaa'a*,
So that your offspring will never be in want.
O Ku-kalaula,
O four thousand gods,
O forty thousand gods,
O four hundred thousand gods,
Take away my sickness,
The sleeplessness, the longing,
The yearning for Holualoa.
Here is my offering—
A leaping *pao'o* fish.

" Then throw the *pao'o* fish into the sea and do not look back. Come straight home. By that time your parents' small *imus* of potatoes will be done. There are three of you, and the small potatoes of the *kala* variety. The *kala* is to free you from thinking of and remembering the handsome man, Holualoa."

To help in inducing pregnancy. (75.8) :

Take two *hinālea* fish, wrap in *ti* leaves and cook on the coals. The first fish is Ku's, and before eating it pray, thus :—

" O Ku, (mention the name of the woman) is going to eat the *hinālea* fish, an offering to you, O Ku. Grant us a child, an offspring of yours in this world . . ."

If a son is wanted, mention the duties of a man, such as fishing or house building. If a daughter is desired, mention the duties of a woman, such as *tapa* making, plaiting, or other things. The second *hinālea* fish is Hina's, and the *kahuna* uses the same prayer, mentioned above.

It was necessary to take care of bones as well as uncon-
sumed flesh of offerings. Mrs. Pukui says, " Each time
students graduated from a *hula* school, or course of lessons,
each ate a whole *'ama'ama,* or as much of it as possible, and
took care to throw the bones into the sea, lest they be defiled
by someone's walking over them.

AUGURIES AND OMENS

Auguries were sometimes made from foods.   An
instance—" Two runners then race.   Properties on both
sides were wagered . . . the priests perform their auguries
with pigs, chickens, and red fish." (25, Vol. III:198.)

When abnormally large schools of certain fish, especially
the *alalauwā* and *moi* and *uiui* were seen, they were regarded
as omens of some unusual event, sometimes only a change of
power among the high chiefs, but usually death.   In an
article on fishing (75.16), this is stated:—

> From olden times down to the time of Keelikolani, the
> appearance of the *alalauwā* is said to be the sign of the death of
> a chief.   We Hawaiians know that when the *alalauwā* appear
> in schools, men and women are sure to ask, " What chief is going
> to die?   Who is it to be?"   As soon as the *alalauwā* season is
> over, the sad news is heard that a chief is dead, and grief comes
> to the people from Hawaii to Kauai.

Mrs. Pukui adds that such schools of fish appeared prior to
the deaths of both Kaahumanu and Nahienaena, wife and
daughter of Kamehameha I. Mrs. Albert F. Judd adds the
names of Liluokalani, in 1917, and Kuhio, in 1921, both of
them descendants of chiefly families.

USE OF FISH BY MEDICAL KAHUNAS

Hawaiian medical *kahunas* prescribed some articles of
food as a *pani* (closure) after a period of treatment.   This
was commonly sea food, though not always.   Handy describes
this custom (35, p. 25) and gives a few prescriptions (p. 22)
wherein *'a'ama* crab and a " white-fleshed fish " are required
as *pani.*

DISTRIBUTION AND MARKING OF FISH BY THE GODS

The gods evidently managed the distribution of fish in
the sea.   A note of this power is in the story of Lonoikamakahiki (25, Vol. I:296) :—

> It was a well-known fact that no sharks were caught on
> these fishing grounds, as the place was dedicated to the gods . . .

the gods had charge of the place; but by the supernatural powers of Loli and Hauna, the fishing grounds known to be without sharks became a place infested with them.

A clearer evidence of the part of the gods in distributing fish is in the story of Maikoha (25, II:270-272). Maikoha was driven from home for his misdemeanours, and was followed by his sisters. One " was changed into that fishpond in which mullet are kept and fattened . . ." at Honouliuli, Ewa. Another married a man named Kaena, a chief of Waianae. " She changed into that fishing ground directly out from Kaena Point and the fishes that came with her were the *ulua*, the *kahala* and the *mahimahi*." Another sister continued on to Waialua, where she married. " The fish that accompanied her from her home was the *aholehole*. A fourth sister went on to Laie, married a man named Laniloa. " The fish that came with her was the mullet and it too remained there to this day."

There is an interesting explanation of how the fish became marked in an old contribution by S. M. Kamakau in 1845, when he was at the missionary seminary at Lahainaluna, on Maui. (75.50) :—

> These islands of Hawaii were created by Kumuhonua (whose wife was Haloiho). He slept and when he awoke, the earth turned and this was called an earthquake.
>
> At that time the duty of each creature had not been apportioned, nor were names given to each . . . So all things were gathered together—animals, birds, crawling things, winged things that fly through the air, and man. The work of each was assigned . . . It was at Molea in Hamakua that all the fishes gathered, the big fish and the little fish. It was there that all the fishes were marked, and streaked ones, the red ones, the white ones, the yellow ones and all the kinds found in the ocean. Kapuhili was the overseer who marked them. The unmarked fish were spotted simply by having ashes sprinkled over them. Then the proper names were given to each variety of all the fishes of the ocean.

HAWAIIAN NOMENCLATURE

A great many Hawaiian names of fishes have survived, even though much of the rest of Hawaian knowledge of fishes—as habits and methods of capture—is now unknown to Hawaiians. The introduction of equivalent names from other parts of the world has been avoided almost entirely. Even among Japanese fishermen who have equivalents for a great many Hawaiian fish that are common to the whole Pacific, Hawaiian names are those most often used. (Hosaka, per. comm.)

Names have survived through constant use. There has been no break in the use of fish as food. And a second reason for survival is because knowledge was held in high esteem, and experts trained pupils in all their knowledge, which included names.

Not all fish are found in any one locality. Even today, with greater ease in transportation, good Hawaiian fishermen are ignorant of some of the less common fish not found in their own marine territory. What the technique of teaching was we do not know, except that there were memory aids. One such aid has survived in the form of a chant listing the creatures of the sea. Mrs. Pukui has made a record of this chant, a few lines of which are given. All of these creatures are of the sea, but not all of them are fish.

> 'A'ama and 'a'aula
> 'A'awa and also the 'aha
> Then the 'aha huluhulu
> And again the ahuluhulu.
> Then the 'akilolo
> And the ala group—
> Ala'ihi, alalauā,
> Alamo'o, alealea,
> Alo'ilo'i, 'ama'ama . . .

In spite of a conscious effort to hand down knowledge, names for some fish did vary from island to island, and even from one part of an island to another. This may be due in part to faults of memory, though memories were trained to astonishing capacity, and in part to a conscious wish to call a fish by what seemed a more appropriate name. Whether the process of change in names was more rapid after European contact than before cannot be known.

All fish were called i'a by Hawaiians, but this term was also applicable to all marine fauna—vertebrate and invertebrate. Not all fish were given definite names. Those that were utterly useless as food and of no importance in any other way were referred to merely as i'a.

Hawaiians gave two names to most fish, one designating the kind (or species), the other designating a group characteristic. The names usually chosen were descriptive of the colour, structure or habitat of the fish. For instance, the *Balistidae*, or trigger-fish, are *humuhumu*. Names of nine *humuhumu* have been collected, as *humuhumu 'ele'ele* (black), *h. mimi* (malodorous), *h. nukunukuapua'a* (nose like a pig), and others. Another grouping is of the *'o'opu*, or gobies. Some were *'o'opu kai*, or sea *'o'opu*, others *'o'opu wai*, fresh-water *'o'opu*. Each *'o'opu* had its own name also. Another distinction was based on habit. For example, *i'a o ke ko'a* referred to reef fishes and *i'a o ke kai uli* to fish of the deep sea. Another grouping included all fish with long sharp beaks, the *a'u* fishes—sailfish, marlin, swordfish, garfish, half-beaks. Each had its distinctive or specific name as well. One meaning of the term *a'u* is to prod. To fishermen in canoes, sometimes far out at sea, the strong beak of all these fishes was the character of most importance. A swordfish or marlin could pierce a canoe.

Though Hawaiians knew fish very well in their outward form, and enough of their organs to name the principal ones, and knew a great deal about their habits, they did not display any knowledge of family relationships other than to bind together by names the fish displaying obviously and easily seen characteristics. Malo (63, pp. 70-73), who wrote at the instance of missionary teachers, lists many sea creatures, fish and invertebrates, and groups them under such classifications as those that have feet with prongs, those that are " beset with spines," those that are covered with heavy shells, those that move slowly, the small fry seen along the shore, those with eminences or sharp protuberances, those that have flattened bodies, and those that are " greatly flattened," those of a silvery colour, those with long bodies, those of red colour, those furnished with rays or arms, those that breathe on the surface of the water, and those provided with long fins or wings. Groupings by such criteria cannot have been of any use to the Hawaiians, except as an exercise for the mind, a play, unless it were to assist memory.

As stated above, Hawaiians gave names to fish for some prominent characteristic of colour, form, or habit. Examples of these terms are (1) colour: *lelo* (reddish), *mele* (yellow), *uli* or *uliuli* (blue, also means green, dark-coloured, that is, the colour of the deep blue sea), *kahauli* (dark-striped), *kea* (white) ; (2) form: *po'onui* (large-headed), *waha nui* (large mouthed) ; (3) a special characteristic: *makaonaona* (bright-eyed) ; *moe* (sleeping), *holo* (travelling), *ka'aka'a la'au* (stick rolling), *pili ko'a* (coral clinging).

For many fish there were—and they still survive among the Hawaiian-trained fishermen—distinctive names for stages of growth. For example, when the mullet (*'ama'ama*) is at finger length it is called *pua 'ama'ama* (or one of these names: *pua 'ama, pua 'o'olola, pua i'i*) ; *kahaha* at hand length; *'ama'ama* at eight inches or so; and *'anae* when fully grown—twelve inches or more. *Pua* is a term meaning blossom, or flower, and is therefore a diminutive for several fish—*pua awa*, etc. Mullet is usually referred to as *'ama'ama,* the age at which it is most delicious, rather than *'anae,* its adult stage. Because fish spawn is similar in appearance a single term was used to indicate the spawn of many fish— *'ōhua,* or *'ahua,* the full meaning of which is given under *'ōhua* in the descriptive list of fishes, herein.

There are instances of Hawaiians disagreeing among themselves as to the proper affinities of a fish. I have been able to find three such recorded instances. There was disagreement, it appears, as to the remora, or shark-sucker. There were two names for this fish, *leleiona* (to dash about in a frivolous way, now here, now there) and *keiki o ka manō* (child of the shark). Kepelino (52) says: " Some say it is not a shark, but it is." Another error concerns the young of the eel (*puhi,* a term which covered eels and eel-like fishes). It is thought by some to be the child of the sea cucumber, because a slender fish *Encheliophis* (*Jordanicus*) *gracilis,* evidently classed as an eel by Hawaiians, is often found in the canal of the sea cucumber, being commensal with it. (20, p. 84.) Still a third misconception was that the *hāpu'upu'u,* a sea bass, was the adult form of the *okuhekuhe,* one of the fresh-water gobies, because of the similarity in head form. It was believed that the young remained in the streams, the adults travelled out to deep water.

Some names of fishes show the relationship of Hawaiians to other Polynesians, and are therefore very old. But many names are peculiar to Hawaii. Translations have been furnished by Mrs. Pukui for all names that seem to have a meaning. For some the meaning is lost or perhaps they have always been names only, without other meaning. The

aptness of the names is to be remarked. Such a chant as the
following fisherman's prayer shows how the composers of
chants preserved observations of fish characteristics.

> To the *aumakua* from the east to the west
> To those of the upland, those of the shore, to Ku-ula, to Aiai.
> To   Hina-who-produces-fish,   and   to   Hina-who-produces-plant
>   foods,
> Eat ye that which I give
> At Hamakua of the steep trail
> That winds up the precipice.
> Blow luck for the fisherman.
> That he may catch the gumless *uhu* at sea,
> The eel with irregular teeth,
> To bring back to the stay-at-home
> The big mullet that listens to loud noises,
> The little *olola* that sleeps on the water,
> The *uauoa* fish that lives near the shore,
> The great octopus of Haaluea,
> The bright-eyed *kole* that dwells in holes,
> Swimming about with the stripe-skinned *manini*
> And the gliding *hou-kahi*.
> May he catch the great *kumu* (coloured like) the *lehua*,
> The red-skinned *awela*,
> The little, swift *kalekale*
> The swimming *a'u*, the *'o'oi* fish,
> The good *uku* fish,
> The *umaumalei*, chief of fish,
> The *oeoe* fish of the deep,
> The *aku* fish that makes a rumbling noise,
> The glistening *opelu* fish,
> The *malolo* that flies over the rising wave,
> Flies with drooping wings,
> The *puhiki'i* (flying) crookedly,
> The *iheihe* fish, too,
> The *a'u* fish with a long snout,
> And the strong gilled *'ahi* of Kalae.
> Fill the *lauhala* (baskets) to carry on the back.
> Dig profoundly, dig deeply,
> Grant life to us on this earth.
> Our prayer is freed.   (75.49.)

In the following descriptive list (pp. 54-139) all names
have been included whether the fishes were used as food or
not, and whether the fish is now known or not, as the name
alone may recall the fish to Hawaiians not available for con-
sultation. The list is derived from Jordan and Evermann
(40), Cobb (13, 14), Malo (63), Bryan (11), the Andrews'
dictionary (2), Kepelino (52), manuscript compilations of or
once in the possession of Kalakaua, Liliuokalani, Hyde,

Henriques and Bryan, and three anonymous manuscript compilations printed in the newspapers (59, f, g, h). A few names, notably those of young stages of growth, have been furnished by Hawaiians of today, many of them by Pa'ahana Wiggin. Most of the 400 or more casts in Bishop Museum bear Hawaiian as well as scientific names and this data has been used. Some names have been found in legends and chants.

The most extensive work on Hawaiian fishes is that of Jordan and Evermann (40). Their collections were made during the summer months, from various places, some of which they specify, and add " and other places " (*ibid.*, p. 20) including the " excellent fish market at Honolulu, which furnished the richest and largest part of the collection." No other work competes with theirs in setting down Hawaiian names.

The casts in Bishop Museum were made by John W. Thompson who worked at Bishop Museum as artist and modeller from 1901 to 1928. He developed many of his techniques of preparation, and the colours are bright and true to this day. In the report of the Director of Bishop Museum for 1901 (10, p. 5), Dr. Brigham says: " Mr. Thompson's work received the emphatic approval of the gentlemen of the United States Fish Commission, and I doubt if so good representations of fish can be found in any museum."

All scientific names have been checked with Fowler's list (26), his work being later than that of Jordan and Evermann. On the subject of scientific names, Pietschmann says (72, p. 6) :—

> Concerning the description of new species, I emphasize the fact that in some groups . . . it is difficult to decide whether or not a species should be separated as a new one from those already described. Many descriptions of known species in these groups are based on a few or even a single specimen, so individual variation and differences in age or sex have not been considered at all.

Descriptions of fish by Hawaiians are few and vague, wholly inadequate for scientific requirements, and as apt to be concerned with taste as appearance. In the descriptions which have been collected from written records, such comments are frequent as " the *kawakawa* is like the *aku*." According to the account of one writer, Kepelino, almost all are " delicious." Colours and markings of fish change so much from youth to age, and colours change so violently when fish are taken from the water that many seeming contradictions in data are nevertheless all true for the moment of observation. Size of fish is a variable character, and all figures must be taken as indication only. A few notes on habitat, behaviour and characteristics of movement have been included when such information was available merely to aid recognition, rather than as an attempt at fully covering such subjects. Sketches have been made chiefly after illustrations in Jordan and Evermann (40). A few are from casts in Bishop Museum, and a few from other sources.

There are a few Hawaiian fishermen still living who were trained in the Hawaiian knowledge. An ichthyologist, working with them, could doubtless glean much additional Hawaiian knowledge of the fishes in these waters. In a few years the opportunity to get the Hawaiian contribution will be gone, only a fraction of it being available now.

Anatomical Terms

Hawaiians give names to all parts of fish of interest to them, as follows:—

1. *nuku,* or *nukunuku*: nose.
2. *lae*: frontal region over eye.
3. *alo*: chest; *alo piko*: belly.
4. *mahamaha*: gill plate.
5. *api*: gill opening.
6. *pihapiha*: gills.
7. *halo*: gill fin.
8. *kualā*: dorsal fin (no separate name for soft dorsal).
9. *kualā lalo*: ventral fin; anal fin.
10. *unahi*: scales; *unahi kalakala*: the rough scales from midbody to tail of certain fishes—scutes.
11. *kakala*: knife-like cartilage near the tail, as in the surgeon fishes.
12. *hiʻu*: tail.
13. *pewa*: tail fin.
14. *umiumi*: barbels (same term as is used for beard of a man).
15. *kiwi*: the " unicorn " of the *kala* fish. (Pukui, per. comm.)

Terms for stages of growth:*

1. *aka*: body is still transparent.
2. *hāuli*: body darkens.
3. *mana*: markings appear.
4. *kakau*: fully marked, fullest stage of development.

* No fish was ever called by any of these terms; some fish did have special names at these stages of development.

LIST OF FISHES BY SCIENTIFIC FAMILY NAMES.

Acanthuridae	api, kala, kole, lai-pala, maiko, manini, naenae, paku'iku'i, pualu, palani.
Albulidae	'o'io
Alopiidae	manō
Atherininae	'iāo
Aulostomidae	nūnū
Balistidae	humuhumu, 'ui'ui (?)
Belonidae	'aha, a'ua'u
Bramidae	mukau
Brotulidae	palahoano
Carangidae	akule, kahala, lai, omilu, ulua, pa'opa'o.
Cephalacanthidae	lolo-oau, pinao
Chanidae	awa-āua
Chaetodontidae	kihikihi, kikakapu, lauhau, lauwiliwili, nuku-nuku.
Cheilodipteridae	'upapalu
Clupeidae	makiawa
Coryphaenidae	mahimahi
Dasyatidae	hīhīmanu
Diodontidae	kokala
Elopidae	awa
Engraulidae	nehu
Eulamidae	mano
Exocoetidae	malolo, lelepo
Gempylidae	hauliuli, walu
Gobiidae	'o'opu
Hemiramphidae	iheihe
Holocentridae	'ala'ihi, kalakoa, 'u'u
Istiophoridae	a'u fishes (sailfish, marlin).
Isuridae	manō
Khyphosidae	nenue
Kuhliidae	aholehole
Labridae	'akilolo, 'ala'ihi, hilu, hinālea, hou, kupou, laenihi, omaka, opule, palemo, paniholoa, po'ou.
Lutjanidae	ahunihuni, opakapaka, uku, 'ula'ula
Mobulidae	hāhālua
Molidae	kaumakanui, kunehi, apahu, makua.
Monacanthidae	loulu, 'o'ililepa, 'ui'ui (?)
Mugilidae	'ama'ama, uouōa
Mullidae	kūmū, weke, moāno, munu.
Muraenidae	puhi
Myliobatidae	lupe
Ophichthyidae	puhi
Ostraciidae	pahu, moa
Pleuronectidae	pāki'i
Polynemidae	moi
Pomacentridae	alo'ilo'i, a'awa, hanui, kupīpī, mamano.
Priacanthidae	aweoweo

Scaridae	......	...... uhu
Scombridae	......	...... 'ahi, aku, kawakawa, ono, 'opelu
Scorpaenidae		...... po'opa'a, nohu
Serranidae	......	...... hāpu'u
Sparidae	......	...... mū
Sphryaenidae		...... kākū, kaweleā
Sphyrnidae	......	...... manō
Squalidae	......	...... manō
Synodontidae		...... 'ulae, welea
Tetraodontidae		...... 'o'opu hue
Xiphiidae	......	...... a'u (swordfish)
Zanclidae	......	...... kihikihi

DESCRIPTIVE LIST OF HAWAIIAN FISHES.

(In this list, Hawaiian names of fishes are usually abbreviated to one letter when names of varieties are mentioned, as *manō*, and then *m. luia*. This is merely a space-saving device.)

'A, or *'a'a* (bright), see *alo'ilo'i*; may also be spelled *'a-lo'ilo'i*.

'A'akimakau (bite the hook), a small fish, a bait-nibbler, such as is the *'alo'ilo'i*.

No data received describing a specific fish of this name.

'A'alaihau (shaped like *hau* leaf).

No data, perhaps a local name for one of the chaetodons, the name suggesting *hau* leaf.

'A'alaihi, the young of the *'Ala'ihi.*

A'awa, certain wrasse fishes: *Lepidaplois bilunulatus* (Lacépède) ; *L. modestus* (Garrett).

Hawaiian names of varieties (yellow) : *a'awa leloa, a. uleholu* (a yellowish-green variety of sugar cane), (island of Hawaii) or *a. e'a,* or *a. hai e'a* (Kauai). Watson says the colours of *a'awa* change to paler tints if the bait is not agreeable.*

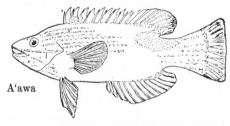

A'awa

Description: *a'awa lelo* (probably *L. bilunulatus*), reddish-yellow, deeper colour than in Jordan and Evermann (40, pl. 24). (In Kan'eohe this fish is called *pō'ou.*) A large dark spot at the end of the dorsal fin, and a smaller spot at the base of the second and third

* Note: Watson adds that colours of all fish change when they are startled, such as by hitting a net, etc.

spines of the dorsal fin are characteristic. The upper part of the head is heavily streaked or blotched with dark brown, changing to dots on the cheek: *a'awa e'a*, or *hai e'a* or *uleholu* are probably *L. modestus*, cast (No. 276) of which in Bishop Museum is dark blue-grey, mottled, with rosy colours in the central portion of each side, including the pectoral fin, other fins blue-grey. Scales large, thin, vary in size; teeth strong, 4 large teeth in front of each jaw; length, about 16 inches; the *a. lelo* is the smaller variety.

A common fish of the coral reefs at all times of the year.

Eaten broiled or dried, can be used in other ways; flesh is white. Sometimes taken as a *pūpū* (bit of food taken after drinking *'awa*, as an aftertaste).

Aeaea (comes to surface, goes down, comes to surface, etc.).

Abundant at Wäihe'e, Maui. A small green fish resembling the *hinālea;* scales soft. Common in Ka'u district, Hawaii, inshore, sea pools and deeper water. Eaten raw or cooked in *ti* leaves, broiled or baked. Used as *pani* (food or drink taken to finish off a medical treatment) for certain children's diseases of the *ea* type. (Comprehensive term for malnutrition diseases.)

'Aha (cord-like) (*a'u a'a, a'u a'u*) (small *a'u*) or *keke'e* (crooked), needle fish: *Strongylura indica* (Le Sueur), this may be the *a'u a'u* (40, p. 124), *S. appendiculata* (Klünzinger), *Ablennes hians* (Valenciennes), perhaps the *a. uliuli,* because it seems to be the largest, *Belone platyura* Bennett. Hawaiian names of varieties: *'aha mele* (yellow), *'aha holowī* (thin), *'aha uliuli* (dark). Young stage: *'aha'aha.*

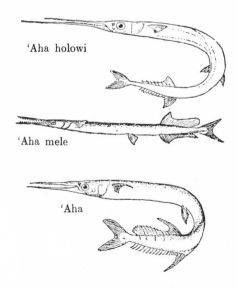

'Aha holowi

'Aha mele

'Aha

Description: *'aha holowī*, about 1 foot to 18 inches; silvery white below, colour of upper part of body described as deep sea-green, and also as blue-black; *'aha mele* has yellow spots (Mrs. Pukui) and *a. uliuli*, about 3 feet, darker in colour; fins of all are greyish. Teeth very sharp. " Both jaws produced in a beak, the lower jaw the longer, very much longer in the young . . . scales small, thin . . ." (40, p. 122).

Common on most reefs, travels in schools near the surface; season continuous; feared because they often leap out of the water when startled and may pierce a fisherman as they descend. *'Aha uliuli* lives in deep water, others in shallow water.

Eaten broiled; flesh is delicious, both flesh and bones slightly bluish when cooked, not harmful.

'Ahi, albacore, yellow fin tuna: *Thynnus thynnus* (Linné), *Neothunnus macropterus* (Schlegel), and others: *Kishinoella rara* (Kishiniuye); *Germo alalunga* (Gmelin), *Parathunnus sibi* (Schlegel), *Thunnus orientalis* Schlegel. Hawaiian names of varieties: *kahauli* (dark striped), small; long pectoral fin, reaches about 75-80 pounds weight; *kihikihi* (angular), long body, long pectoral fins; also called *'ahi 'opūhemo* (stomach disgorged when it is brought to the surface), 60-80 fathoms is the depth to fish for it; *mahao'o* (strong-gilled); *malailena* (yellow fins), long body; *maoli* (indigenous), large, about 200 pounds, flesh light, slightly pinkish in colour; *palaha* (flattened), *Germo alalunga* (Gmelin), a small sized *'ahi*, perhaps a stage of growth; *po'onui* (large head), short body and thin tail are characteristic; flesh is fat and delicious; *hi'u-wīwī* (thin-tailed), also called *ka'aka'a-la'au* (stick-rolling), well known for its strength in pulling on the hook, has a tough lip, similar to and possibly the same as the *kahauli*. An early stage of growth: *kananā* (fighter), lighter in colour.

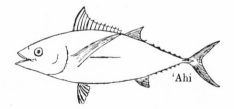

'Ahi

Description: Size, from 75 to 200 pounds, body stout; colour, dark steely blue above, under body silvery; dorsal, anal and all finlets bright yellow, other fins touched with yellow; scales small.

A deep sea fish, Kona and Ka'u districts in Hawaii famous for 'ahi; the chiefs used to go to that coast for the sport of fishing for it; one of the gamest fish. The chief season is May to August.

Eaten raw, baked in the *imu* whole, or cut in pieces, and wrapped in *ti* leaves then baked, or dried. For drying, the fish was first cut into steaks or strips, and sometimes baked in the *imu* before drying in the sun. Left-over fish from a meal was dried. The heavy flesh needed special care. For drying raw fish, the entrails, head and tail were removed, the flesh cut into sections about 6 inches long, sprinkled with salt and put into a tightly woven *lauhala* basket or a calabash (*umeke*), or, in modern times, a barrel, then salt was poured on and the top covered with salt. Care was taken to poke the salt—with a bamboo (*'ohe*) stick—deep into the meat around the bones. After three days the flesh was taken out, rinsed and set to dry in the hot sun, preferably on the flat stones of the beach where the reflected heat would help to dry the under sides. The drying continued several days, sometimes a month. So prepared, 'ahi will last for years in a dry climate. To cook dried 'ahi it was wrapped in *ti* leaves and baked in the *imu* for about three hours, by which time the colour became mellow and the taste enticing. If the 'ahi were in strips, they were sometimes rolled into bales before wrapping in *ti* leaves.

Aholehole (sparkling) (*aholele*, or *ahole*), one of the perch-like fishes: *Kuhlia sandvicensis*. Stages of growth: young, *pua-hole*, or *pua* (descendant) *aholehole*; half-grown, *'apo'apo* (grasp), a term not used everywhere (not in Ka'u); full grown, *aholehole*.

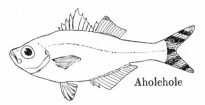
Aholehole

Description: Length sometimes reaches 10 inches, average about 6 inches; colour, silvery grey above, fading to white underpart of body, fins touched with grey.... The skin is fine and transparent, eyes large, scales large. When the fish is fat (in roe?) the underbody takes on a yellowish tinge. Foreigners were sometimes called *aholehole,* in the early days of foreign contact, because of their white skin colour.

A common shore fish, found in both salt and fresh water, the young abundant along sandy beaches, chiefly in shallow water; when mature their habitat is the coral or lava caverns of the reef. They travel in schools. Brock says " There seem to be the reef or coral *aholehole* and the *aholehole* that enters fresh water. *K. marginata* may be the reef species which is characterized by a shorter snout and larger eyes in specimens of comparable size."

Eaten raw, dried, or broiled on hot coals, also salted or *ho'o-melumelu*. In preparing for eating raw, the dorsal fin is removed by making a cut on each side of the fin, and then the fin can be pulled off, Hawaiians doing it by taking a good grip with the teeth and pulling. This is one of the fish that was considered a dainty, and often craved by chiefs. " The chiefess yearned again for the *opule* fish and the fat *aholehole* fish of Wai'akolea (Puna, Hawaii). It is far from Hilo to Puna, but because the chiefess had a craving the distance was as nothing. The *opule* fish was brought to her alive, still breathing, inside of a wrapping of *pakaiea* sea weed. The *aholehole* fish was still moving, wrapped in some *kala* seaweed found in the pond." (75.42.) The *aholehole* was used in sacrifices when a white fish was needed, as in a ceremony to keep away evil spirits.

A man was sent to the beach for an *'ohiki* crab, an *'a'ama* crab and an *ahole* fish. The prophets told the men to bring these things together with *'awa* (a ceremonial drink) to the sand where the canoes ... had departed ... (75.48.)

These creatures were wanted because of the meanings of their names: *ohiki* means to pry loose, *a'ama* means to loosen one's grip, and *hole*, in *ahole*, means to strip away. Thus the evil influence would be pried off and stripped away. Another specific use in magic was in warding off evil influences from a house under construction. *Aholehole*, or the *kole*, were put under the house posts when they were set. If a *kahuna* (priest) should enter and predict trouble after this precaution had been taken, he himself would receive the trouble and would die. (75.2.)

The *aholehole* is one of the " sea pigs " (*pua'a kai*) used as substitute for pig for certain ceremonies, or for any ceremony when pig was not available.

'Ahua (grown larger). This is a term applied to the young of two fish, the *kawakawa* and the *aku*, indistinguishable until they mature, and found together. The spawn is called *kina'u*, the second stage *'ahua*, the adult stage *kawakawa* or *aku*, whichever it may be.

A favourite place to take *kina'u* was at the mouths of the streams of Kauai, where they are said to have fed on the *hala* (pandanus) fruits that floated down the streams. An old saying is, " Surrounded by the net is the *kina'u*, caught by the hook is the *'ahua*, enclosed in the pond are the *kawakawa* and the *aku*." (*Puni i ka 'upena ke kina'u, lou i ka makau ke 'ahua, po'i i ka pa ke kawakawa me ke aku.*)

The young of the *moana* are called *moana 'ahua*. The young of several fishes are called *'ōhua*, the stem, *hua*, meaning egg or seed.

Ahuluhulu. The young of the *kūmū*.

Ahunihuni.

One informant, Nakuina, says this is a fish of the *opakapaka* class, one of the snappers; small, being only 3 to 4 inches long; it lives in the deep sea, 30 to 40 fathoms down, and when raised to the surface the " soft parts " are forced out from the body, as in some other deep sea fishes. The *kahala* and the *uku* feed on this fish.

Ahuula, see *auhu 'ula.*

Akaka (clear), listed (59a).

Akekē (big belly), see *'o'opuhue.*

Akiki (small), same as *ukikiki,* the young of *opakapaka.*

'Akilolo (bite brain), see *hinālea 'akilolo.*

This fish was used as *pani* (closing dose) for any sickness of the head; sugar cane bearing the same name was used for medicine for head sicknesses.

Aku, ocean bonito: *Katsuwonus pelamys* (Linné). Stages of growth: *kina'u* (imperfect, immature), the spawn; *'ahua,* half-grown; *aku,* full-grown.

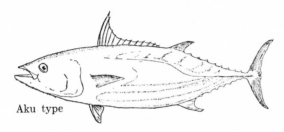

Aku type

Description: Average length, 2 to 2½ feet, body stout; colour is deep blue above, with deep blue stripes below the lateral line; no scales.

Eaten raw, dried, cooked in the *imu,* or used for *palu.* The reddest flesh, next to the bones (*i'o alāea*) was too rich in oil for those with skin trouble to eat. One writer says the head is delicious, " for the eyeballs are there." Throughout Polynesia, the eyeballs are prized as a food delicacy. Kepelino describes most fish as delicious, but says the *aku* is the most delicious, " especially when broiled until dry " (that is, until the juices cease to drip). One informant says, " They eat this fish from head to tail—gills, everything."

One of the best known traditional stories is that of the coming to Hawaii of Pa'ao, a chief of " Kahiki." He quarrelled with his brother, Lonopele, who, by supernatural powers, sent storms after Pa'ao. But a school of *aku* appeared, " rose to the surface with a roaring sound," and by swimming about calmed the water. Lonopele brought on a second storm which threatened to swamp the canoes, and again a school of fish appeared in response to prayers of Pa'ao, this time the *'opelu* fish. The waters were made smooth again, and Pa'ao and his company arrived in Hawaii. To honour these two fish they were made tabu to the families of those who had been in the canoes, Pa'ao and his company, as well as tabu to all Hawaiians for a few days each year. (47.) Vancouver (85, Vol. III: p. 18-19) speaks of this tabu occurring in February and lasting ten days. He also notes the dexterity with which the Hawaiians catch *aku* (85, Vol. II: p. 168) and its great value to them as food. They are " exceedingly good to eat when fresh, and being caught in abundance, make a very considerable part of the food of the inhabitants when preserved and salted."

As with other fish, the first catch was reserved for the offering to the fish god. By a " wicked person " an *aku* was sometimes thrown under the thwart of the canoe, so that it would be defiled by being sat over, unfit for offering and therefore available as food.

Akule, big-eyed, or goggle-eyed scad: *Selar crumenoph-thalmus* (Bloch). Stages of growth: *pa'ā'ā* (strip-ling), 2-3 inches long; *halalū* (or *hahalalū*), 5-6 inches long; *akule*, adult. Informants disagree, some say *halalū* is the youngest stage, *pa'ā'ā* the next older.

Akule

Description: Size, 7-12 inches, occasionally as much as 20 inches in length; colour, dark blue, with brassy reflections, ventral lighter, almost white; caudal and dorsal fins dusky, lateral line very prominent; scales small.

Eaten raw, broiled, cooked in *ti*-leaf bundles placed over the taro in the *imu;* good for *palu;* a favourite fish for drying.

As to its habits, one writer claims that:—

The *akule* is a fish that goes from place to place; if a school of them makes a first appearance at Kahana, it will divide into two groups, one going towards Ko'olau, and the other towards Wai'anae. They often met at Kailua and many were caught. (75.28.)

It is found most abundantly in the big bays of the islands, especially those of Kauai; sometimes called " the fish of Hanalei " (large bay on Kauai). In former years a school would invariably come into the harbour of Honolulu when they were at the *halalū* stage. They were easily caught, one pulled in after another—millions would be taken. When a school was surrounded, a *kahuna* (priest) or *kilo* (watcher) would pick out a first fish, and throw it back into the sea, with a prayer to the local fish god, or Ku'ula himself. *Akule* are most abundant at Kane'ohe in April, May and June.

Akupa (big mouthed), see under *'o'opu.*

'Ala'ihi (or *'ale'ihi*). Certain of the *Holocentrus* species (squirrel fishes) were called *'ala'ihi*, possibly all of them. Hawaiian varieties are: *'ala'ihi kalaloa* (long-spike) (or *kanaloa,** or *kakaloa*), *H. diademus* (Lacépède); *a. maoli, H. xantherythrus* Jordan and Evermann; *a. mahū, a. ako'ako'a, a. piliko'a* (clinging to coral), these unidentified with scientific names. The *'ala'ihi lakea* (white dorsal) has been identified as one of the species of *Myripristis* (soldier fishes), its red colour

* Note: *l* and *n* sometimes changed places in Hawaiian words.

and lengthwise stripes evidently admitting it to Hawaiian relationship. The young are called *'a'ala'ihi*.

'Ala'ihi kalaloa   'Ala'ihi lākea

Description: Usually 2 to 2½ inches in length, sometimes as much as 6 inches; colour of all species bright or rose red, with white stripes, fins red, dorsal often brilliant red, sometimes touched with yellow; it reflects the light. According to Jordan and Evermann (40, p. 158) "remarkable for the development of sharp spines almost everywhere on the surface of the body." Scales large.

Eaten raw, salted, broiled, both fresh and after drying; difficult to prepare because scales are tenacious, "and then two mouthfuls devoured it." *'Ala'ihi* live in holes in the reef in shallow waters, are very common, usually caught at night; said to have been the favourite fish of Kamehameha III (reigned 1825-54).

Alalauwā, young of the *aweoweo;* see under *aweoweo.*

Alamo'o (lizard-like), Hilo name for one of the fresh-water *'o'opu,* see *hi'u-kole* (red-tail), under *'o'opu.*

Alea, or *alealea,* one informant says a short name for *hinālea;* another that it is like the *hinālea luahine* (old woman), but is brown instead of greenish; see *hinālea.*

'Ale'ihi, see *'ala'ihi.*

'Alo'ilo'i (bright and sparkling), a demoiselle fish: *Dascyllus albisella* Gill. Young stage is *'a,* or *'a'a; 'a'akimakau* (bait-nibbling) is a term for a variety, or perhaps an alternative name.

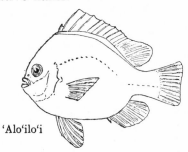

'Alo'ilo'i

Description: Size, reports differ, 5, 6, or 8 inches long; colour, brownish, the large scales are white with black borders, except in the middle dorsal portion of body which is lighter in colour; all fins dark, except for whitish scales on dorsal fin. Colouration varies. Kepelino (52) says " two colours, like mixed paint over the entire body, reddish brown and black in equal degree."

Eaten raw or cooked in hot ashes; a tasty fish, well liked.

'Alukaluka (soft). No data.

'Ama'ama, mullet: *Mugil cephalus* Linné. Young stages: finger length, *pua 'ama'ama,* or one of the following: *pua, po'olā* (young life), *'o'olā;* hand length, *kahaha;* about 8 inches, *'ama'ama;* 12 inches or more, *'anae.*

The *'anae* are said to make seasonal migrations around part of the island of Oahu, from Ewa, south around and up the easterly coast of La'ie, where they remain a few weeks and then return. When migrating they are called *'anae-holo* (running or travelling mullet); when they remain more or less off shore, or have returned from the journey, they are called *'anae-pali* (cliff mullet). Mokumaia (75.34) notes:—

> After the *'anae holo* has returned to Ewa, it is renamed the *'anae pali,* for its appearance has changed . . . An *'anae holo* is light in colour and its body is not so plump. When caught in the net and after ceasing to struggle " perspiration " breaks out all over its body . . . it is as white as a horse that has jumped here and there . . . As to the *'anae pali* . . . its body is plump, resembling the pond mullet. The scales are dark, the gills are red, on both sides of the mouth you will notice traces of red . . . the tail is well-shaped . . ."

'Ama'ama

Description: The young silvery in colour, like money; adult, upper body greyish, becoming whitish only on the undersides; eyes are whitish with dark pupils; a tinge of red at lips and gills; size, about a foot and a half, body robust; scales large.

The mullet is and was the most important fresh water, or brackish water, fish of the Hawaiians. They caught the spawn in nets along the shore, impounded and fattened them in ponds and had a constant supply. It is delicious, the flesh is white and there are few bones to contend with. Eaten both raw and cooked. If raw, a little *manauea* and *'o'olu* seaweed were added. The most popular way to cook mullet was to wrap it in *ti* leaves or ginger leaves, then broil or bake, the fragrance of the leaves being imparted to the flesh. Good cooks were proud to unwrap the fish after cooking and show the fish little changed in appearance, the skin undisturbed. It was also eaten dried. When fat, it was good for *ho'omelumelu.* (See p. 27.)

The legend of the origin of the *'ama'ama* is as follows:—

Kaihuopala'ai (a place) was famous from olden times down to the time when the foreigner ruled Honouliuli, after which time the famous old name was no longer used . . . It is said that in those days the 'ama'ama heard and understood speech, for it was a fish born of a human being, a supernatural fish. These were the keepers of this fish . . . Kaulu, the husband, and Apoka'a, the wife, who bore the children, Laniloa, the son, and Awawalei, the daughter. These two children were born with two other supernatural children, an eel and a young 'ama'ama. From this 'ama'ama child came all the 'ama'ama of Kaihuopala'ai, and thus did it gain renown for its 'ama'ama . . . Laniloa went to La'ie, in Ko'olauloa, and there he married. His sister remained in Honouliuli and married Mokueo, and to them were born the people who owned the 'ama'ama, including the late Mauli'awa and others . . . These were fishermen who knew the art of making the fish multiply and make them come up to the sand . . .

While Laniloa lived in La'ie he heard of the great schools of 'ama'ama at Honouliuli. There were no 'ama'ama, large or small, where he lived. He thought of his younger sister, the 'ama'ama, and guessed that was the reason the place was growing so famous. He said to his wife, " I shall ask my sister to send us some fish for I have a longing for 'ama'ama . . ." Laniloa left La'ie to go to Ewa . . . He reached the house and found his parents and sister. His parents were quite old for he had been away a long time . . . He said, " I have come to my 'ama'ama sister for a bit of fish as there is none where I live except for some au moana (sea-faring) crabs." . . . After three days and nights he left Ewa . . . The fish were divided into two groups, those that were going and those that were staying. As Laniloa's sister went along the shore she went in her human form. The fish came from, that is, left Honouliuli without being seen on the surface. They went deep under water until they passed Ka'a'ali'i, then they rose to the surface . . . They reached Waikiki . . . They went on. The sister slept at Nu'upia while the fish stopped outside of Na Moku Manu . . . Finally she reached La'ie, and to this day this is the route taken by the 'ama'ama. (75.39.)

Another version of this story is credited to Keliipio (49:112-113). The author records that this half-tour and return is made commencing " some time in October and ending in March or April." Dakin (17:17) says that the " sea mullet of New South Wales migrated every winter out from brackish waters of the estuaries to ocean water of higher salinity to spawn." Mrs. Pukui comments that these migrating fish do make frequent stops which are taken advantage of by the fishermen. The fish come close inshore, many times within the breakers.

The popularity of the mullet is attested by the frequency of its being mentioned in tales and legends. A few notes from scattered sources follow:—

From the chief's favourite fishpond, Mapunapuna, mullet were plump and soft when eaten with the poi from our taro patch.

Where did the mullet spawn come from? From the chief's favourite pond, Kahikapu; the fish were small but delicious when bitten into. (75.34.)

Ke'ehi, near Moanalua, was famous for the deliciousness of its 'ama'ama spawn . . . those living on the beach of Ke'ehi left them undisturbed until they reached the *kahaha* stage. (75.28.)

Molokai, island of Hina, had many '*ama'ama* ponds—Ualapua, Ni'aupala, Pi'o-p'io, Puko'o, and Kepeke, noted for the fatness of the '*ama'ama*. (75.11.)

We went to the beach of *Iloli* . . . fish were caught with the feet (meaning very easily) . . . the *pua*, the '*ama'ama* spawn. (75.36.)

Waipi'o (Hawai'i), here lived Liloa who loved the valley for its riches . . . '*ama'ama* among other things. (75.9.)

Mrs. Pukui says that the best liked of all " sea pigs " were the mullet and the *kūmū*.

Amo'omo'o (not up to full growth), young stage of the '*o'io*.

Amuka, see *kahala*.

'*Anae*, adult stage of '*ama'ama*.

'*Ananalu* (belonging to the surf). Only one reference, " We are dazzled by the emerald and purple tints of the '*ananalu*." (75.26.)

'*A-niho-loa* (long-toothed '*a*).

Mrs. Wiggins says it was a small fish, about 8 inches long, a narrow body, and very sharp teeth, difficult to scale, the scales being slimy; not especially good for eating; found at Oahu, not found in Ka'u. In an article concerning the island of Lana'i, there is this statement:—

Just glance at the '*a-niho-loa* with its transverse bars of red and white on the tail, then a body of the loveliest azure, a perfect tricolour . . . (75.26).

'*Ao'aonui* (big sided), the young or an alternate name of the *kupipi*.

Apahu (cut off), see *kunehi*.

'*Api*, one of the surgeon fishes: *Acanthurus guttatus* (Schneider). Another Hawaiian name is *hapi*; perhaps that is the original form of the name. The name '*api* was given to different fish in different districts, such as *Zebrasoma flavescens*, or *Z. veliferum*.

Description: Size 8-11 inches (Mrs. Wiggins says " forearm length "); colour, brown with round white dots and bands; skin tough, scales small; caudal spine small, short, depressible in a groove; body compressed. (40, p. 392-393.)

Delicious broiled, also made into *i'a hoomelumelu*; eaten in any manner.

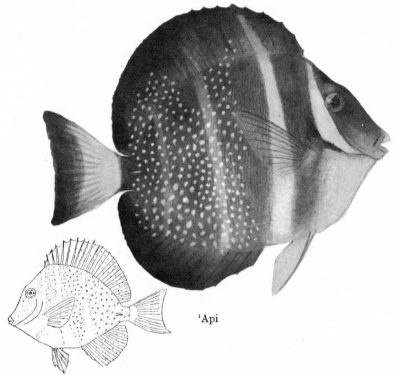

'Api

Lives at outer edges of the reefs; fairly scarce; not known today in Kane'ohe or La'ie; there are plenty in Hilo, Ka'u and Puna.

Apohā (bubble maker), one of the *'o'opu*.

Apu'upu'u, see *hapu'u* (rough skinned).

A'u (prod), swordfish, sailfish, marlin, spearfish: *Istiophoridae*. Information is incomplete.

There is some uncertainty as to Hawaiian species names of the *a'u* fish. Besides the great, deep sea *a'u*, the needle fish, *'aha*, and garfish, *iheihe*, were given *a'u* as one of their names. They were usually referred to by their own names, however, and all their names are included under the most common one. Similarly, the *kākū* was sometimes called the *a'u kākū*. I believe that this linkage in names of fish so different in other respects signifies only that Hawaiians considered the prolonged jaw, or nose the most striking character, the one by which it should be named.

Two names were collected which remain unattached to fish, no informant knowing what they are: *a'u ku'au-lepa* (*ku'au*, handle; *lepa*, hanging piece of tapa), possibly an alternate name for *iheihe*, and *a'u papa'ohe* (board of *'ohe* wood).

The true *a'u* are as follows: *a'u kū* (beaked *a'u*, the broadbill swordfish), *Xiphias gladius* Linné.

A'u kū

Description: There is a cast of a small specimen in Bishop Museum, acquired a few years ago. The colouration is dark grey, almost black over the whole body, except the ventral part, which is light grey. LaMonte and Marcy (52, p. 7, 1941) give the colour as "bronze or purplish-black, whitish below." Other notes from their description are that the world's record fish is 13 ft. 9 in. long, and the fish probably reaches a weight of a thousand pounds or more. One at the Honolulu fish market in 1916 weighed 735 pounds. (24, p. 89.) The fish is scaleless, has no teeth, and, unlike the marlins and spearfish the dorsal does not fold into a groove. In the same paper (55, p. 9), there is a note from Hawaii, sent by Mr. C. M. Cooke, III, stating that young swordfish from two to four feet and less are rather common, caught from three to twenty miles off shore at six to thirty fathoms. Jordan and Evermann (40, p. 168), state that the flesh is red, rich in colour, highly valued as food. This is at variance with a Hawaiian informant who says the flesh is white, and eaten by Hawaiians only when other fish may not be had; the flavour lacks zest for them. A'u kū, as well as other a'u fish, are fairly plentiful off Kailua, Kona; plentiful in August at Kane'ohe, Oahu, says one informant. Because of the gaminess of all swordfish they take too much commercial fishing time to capture—another count against them as a good food fish. They have become sought after in Hawaii as a sport fish of late years.

Though not craved for food, they were eaten when caught, cooked in the *imu*, salted, or wrapped in *ti* leaves and baked in the *imu*. Some say they were eaten raw, but a Ka'u informant says not in her district.

There is a notice in an 1870 newspaper (75.13) about a swordfish that came to the shore near the Russian Consul's house. Its length was 1¾ fathoms (estimated or measured?), more than five feet around the body, and the nose 3 feet long. It was seen by men in a fishing fleet outside the reef. They let down a line but the swordfish headed directly for shore, and a boy ran down and tried to lassoo it . . . The writer goes on:—

> The superstition regarding this fish in the olden days was that it made whoever captured it a conqueror, gave him power to succeed against an opponent. The fish could belong only to a chief, and was laid on the atlar with a human sacrifice, just as was done with tiger sharks. These two fish were both caught with human bait, and both offered with prayers by the *kahuna* for power for the chief . . . When a fish of this kind came ashore, of its own accord, it was an omen that a person closely related to Ku'ula (the god of fishermen) was about to die . . .

A'u lepe (comb of a cock), the sailfish, *Istiophorus.*

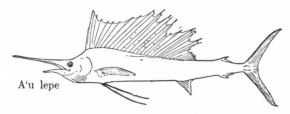

A'u lepe

Description: Deep blue in colour, lighter underneath; a specimen in Bishop Museum is 7 feet long; fishermen say some reach 9 feet. From the original description it is noted that the teeth are fine and sharp, the skin rough, and with fine scales, the dorsal fin depressible in a groove; colour evidently more greyish-blue than in Bishop Museum cast No. 354; there are 17 vertical rows of dirty white spots on the body entirely missing in the cast (possibly they faded before the fish reached the preparator), and the dorsal fin has a multitude of black spots instead of a few present in the Bishop Museum cast.

A'u?; possibly *a'u kī* (thin), marlin, *Makaira* species.

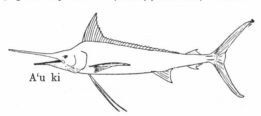

A'u ki

According to LaMonte and Marcy (55, pp. 19-20), the striped, black, and silver are the three marlins in the Pacific. They compare them as to height of dorsal fin and weight at different lengths. A sport fisherman, S. Kip Farrington, Jr. (23, pp. 179-80) says: " The silver marlin is practically the only marlin that is caught in Hawaiian waters . . . Most silver marlin caught off the Hawaiian Islands by the rod-and-reel fishermen run between 150 and 250 pounds."

Thomas Maunupau says that all *a'u* fish live at considerable depths, as all fishermen know, and come to the surface only when the school they are following comes to the surface. Anyone fishing for *a'u* for food would use a deep line, not rod and reel.

A'u? spearfish, *Tetrapturus* species.

According to LaMonte and Marcy (55, 21-22) :—

The position of the genus *Tetrapturus*, the so-called Spearfish, is undecided, and present evidence seems to the authors insufficient for definite opinion. The question is whether there is really a Spearfish, separate from the Marlins, or whether this fish is the young of some Marlin . . .

All these *a'u* or prodding fish were feared for their long sharp, heavy noses, which sometimes pierced or broke the fisherman's canoe.

Aualaliha, a kind of *'o'opu.*

Au'au kī (thin), the young of eels, *puhi.*

Awa-'aua, tenpounders: *Elops hawaiiensis* Regan.

Awa-'aua

Description: 3 to 5 feet long, body compressed; colour silvery, scales thin, large, silvery.

These large fish of the open seas are great game fishes, belonging to the tarpon family. They were also bred in ponds; those in the open sea became larger, but those from ponds, 12 to 14 inches long, were fatter. Evidence varies as to food value. The open sea *awa* some call dry and bony, not much valued. But the pond *awa* are good eating. J. and E. (40, p. 6) quote from the *Blossom* voyage: " A fish which is kept and reared in the taro ponds and esteemed very highly by the natives, especially the belly part, soaked in salted water and eaten raw . . ." *Awa* were sometimes served with seaweed, such as *manauea* or *'o'olu.* In one tale is mentioned the " very fat *awa* of the pond of Poki." (75.17.)

The Chinese of Hawaii are fond of this fish for fish-cake. They let it stand on ice one day, then scrape the flesh from the bones, make it into balls, dip in egg and fry in deep fat.

Awa are common about the reefs, and are usually caught in nets. Beckley says (4, p. 12) that ponds were stocked with them by catching schools of the young in shallow water in nets. They travelled in schools in deep water. " When a school was caught there was a canoe-load . . ."

Awa, the milk fish: *Chanos chanos* (Forskal). Stages of growth: *puawa, awa-aua.* J. and E. (40, p. 57) says the young are *puawa,* medium size, *awa-aua* (describes the state after the bloom of youth is gone; a little rough) ordinary commercial size, *awa;* very large are *awa kalamoho.*

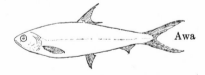

Awa

Description: Size averages 12 to 14 inches; colour, silvery-grey, whitish below, scales rather large.

The flesh is white, delicious. It vied with the *'ama'ama* in poularity. The chiefs delighted in well fattened *awa'āua,* and reserved them for their own use, if the supply was less than abundant. They are mentioned in Hawaiian tales in such remarks as: " Where did the *awa'āua* come from. From the chief's favourite pond, Kaloaloa (near Moanalua) . . . It slipped down the throat with some *o'olu* seaweed. Ah, it was so good . . ." (75.34.)

Awala, or *awela,* see *hou.*

Aweoweo (glowing red), sometimes called big-eye: *Priacanthus* species; *P. boops* (Schneider); *P. cruentatus* (Lacépède); *P. maracanthus* Cuvier, and *P. hamrur* (Forskal). The young are called *alalauwā.*

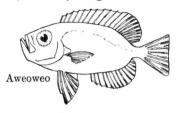

Aweoweo

Description: Colour is brilliant carmine or reddish bronze. *P. boops* is a large variety, 20 inches long, and is oronzy, fins brilliant red except ventral, the membrane of which is black; *P. cruentatus* is smaller, about 12 inches long, brilliant red entirely. J. and E. (40, pp. 228-232) describe some as silvery, but Mrs. Pukui says she is unfamiliar with silvery *aweoweo*—all she knows of are reddish. Scales are fine, sharp and brittle, skin thick and tough. Scales small.

The flesh is white, good for broiling and drying, not good for cooking in *ti* leaves unless skin is removed; not eaten raw in Ka'u, but liked raw in Maui and Kauai; if dried usually eaten without cooking.

Abundant in the shoal waters inside reefs. This fish comes in great schools occasionally. In former years Hawaiians placed superstitious signficance on the advent of a school. Though it brought an abundant supply of delicious food a school inshore or in a harbour was regarded with great awe and sorrow, portending as it did the death of a high chief. One record of a large school coming into Honolulu harbour is from 1873. Between two and three thousand people were out with hook and line. " As soon as the hook and bait could be cast into the water the fish would bite and be hauled in. They were mostly very small, varying from an ounce or two to half a pound." (75.22.) Another larger catch in 1919 was recorded:—

> In former years few foreigners were seen fishing when these fish appeared. Now there are more foreigners than natives on the wharf, thousands in a single night. From rich to poor, all are fishing, because the fish in the market are expensive . . . Hawaiians should continue to fish, for we do not know when the fish will stop coming in. Hawaiians dry all they can, but foreigners sell most of theirs, and reserve only enough for a meal. (75.12.)

There is a note of a large school coming in at La'ie in 1917: " The poles of the women were seen to bend; they say that the *alalaua* caught with hooks taste better than those caught with Logana's (a fisherman of the locality) net ..." (75.38.)

E, listed (59 i).

E'a, see *a'awa.*

Ehe'ula (red *che*), adult stage of *he pu'u,* "a flying fish, *Exocetus,*" says Wetmore (88, p. 95).

'Ehu (reddish), Hawaiian list (59 g).

Description: Colour red; size, about 1 ft. long; a deep water fish, caught in deeper spots than *ulua kahala,* and *ono.*

Eaten raw, but liked better cooked in the *imu,* both fresh and dried, wonderful flavour.

Very scarce, though found about all the islands.

Enenue, see *nenue.*

Hāhālua (divided feeler), sea devil, manta devilfish; *Manta birostris* (Walbaum). Not eaten.

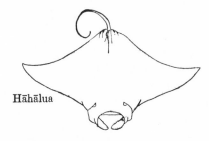

Hāhālua

Halalū, see *akule.*

Halalua, another name for *kaweleā.*

Hahili, see *nohu.*

Haie'a, see *a'awa.*

Hailepo (ash coloured), perhaps *hailipo* (dark coloured), like a *pualu.*

Halahala, young stage of *kahala.*

Halaloa, listed (59 d).

Hanui (thick-set), or *mokumokuhānui.* The dictionary states "same class as *maomao,*" which makes it a demoiselle.

Description: Size, "forearm length"; colour, brown, like *kalekale,* has a big mouth.

May be eaten raw or cooked in any way.

This is a shallow water fish, plentiful in the early summer; seen at Pu'uloa (Pearl Harbour); not known at Ka'u (south-west Hawaii, deep-water shore).

Haoma, listed (59 a).

Hapu'u (rough skin), grouper: *Epinephelus quernus* Seale. Young stage: *hapu'upu'u,* or *apu'upu'u.*

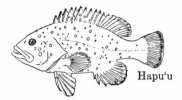

Hapuʻu

Description: About 2 feet long, body moderately compressed; colour, according to cast 394, in Bishop Museum is nut brown (adult colour). Kamakau (47, Chap. 4, p. 38) speaks of the black *hapuʻu* and the reddish *hapuʻu*—perhaps the same fish at different ages. Bishop Museum cast is about 3½ feet long, body deep, robust. Jordan and Evermann (40, pp. 223-224) states that scales are small, rough, and the body has irregular pearly white spots on the sides before maturity; mouth is very large.

A good fish to eat, fleshy, eaten any way except raw; safe to eat in Hawaii, poisonous at certain times in some other parts of the Pacific.

Rather scarce in Honolulu market, say Jordan and Evermann (*ibid*). A deep water fish, 150-200 fathoms is its habitat, therefore few caught. When pulled to the surface, the sudden release from strong pressure eviscerates the fish. Field (24, p. 92) tells of one at Honolulu market that weighed 600 pounds.

Hāuliuli (dark-coloured), sometimes called *hāuliuli puhi* (eel-like *hāuliuli*): *Gempylus serpens* Cuvier.

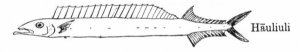

Hāuliuli

Description: From 15 inches to 3 feet long; colour, variously described as greenish-grey, reddish, and dark, metallic blue or black. No scales, slippery to handle, teeth like a dog's, exceedingly sharp.

Flesh is firm, and of a darkish tint; delicious, sweet, one of the best fish to eat. Not eaten raw, but cooked in any way, also dried; few bones (Kinney). In Kaʻu they were liked best dried, toasted a little when prepared for the meal.

Plentiful in some areas, such as Kaʻu, island of Hawaii (deep water), scarce in others, as Kaneʻohe, Oahu. Also found in medium deep water, at outer edge of reef, in early summer.

Kamehameha is credited with saying that *hāuliuli* is a delicious fish for the back country people, meaning good for those who could get no better. Evidently he was not fond of it. There is a figurative expression, " *Liʻiliʻi hāuliuli, monimoni ke ʻae* " (The little *hāuliuli* makes the mouth water), the meaning obvious.

Hehena (crazy), another name for *leleiona*.

Hīhīmanu (bird-like). Notes collected concerning the ray fishes are confused, the names *hīhīmanu* and *lupe* (kite) seeming to change places at various parts of the island group. In some areas the *hīhīmanu* is the sting ray: *Dasyatis brevis* (Garman), and *D. latus* (Garman); and the *lupe* is the eagle ray, or spotted sting ray: *Aetobatus narinari* (Euphrasen).

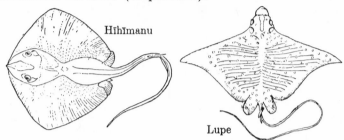

Hīhīmanu

Lupe

Mrs. Pukui believes that neither *hīhīmanu* nor *lupe* were eaten of old by Hawaiians. Oriental peoples love them, but the " one with the long tail " (*Aetobatus ?*) is not good to eat. Occasionally Hawaiians eat them today. They are cut into chunks, salted a few days, wrapped in *ti* leaves and baked in the *imu*; others prepare them like " jug-meat," or " jack-meat," (Hawaiian colloquialisms for jerked meat) strips of flesh salted and dried in tough, hard strips or ropes (*hīhīmanu-kaula*) (Kinney).

They have been noted off the Napali coast of Kauai (deep water, no reef), where they will venture close to shore when a high surf is running. If this is during a calm spell, rough seas may be expected in a few hours. They are common in Kaneʻohe Bay, Oahu (protected waters), in mid-summer, flipping their " wings " along the surface. One *lua* (home base) is said to be off Haleiwa, northern end of Oahu.

Hilu (well behaved) : several varieties among the *Labridae*: *Coris flavovittata* (Bennett), black, banded with yellow, lengthwise stripes; *C. lepomis* Jenkins, *h. ʻeleʻele* (dark), or *h. lauwili* (confusing), green, with blue shades, 12-18 inches long; *C. ballieui* Vaillant and Sauvage, *h. mele mele* (yellow), also called *h. malamalama* (bright, like sunlight). *Malamalama* may have been an alternate name for *hilu* as there are *malamalama ʻula* and *malamalama uli; C. multicolor* (Ruppell), *h. pilikoʻa* (cling to coral). In the *Cirrhitidae* family, the *Paracirrhites forsteri* (Schneider) is also thought to be the *h. pilikoʻa*. Other *hilu* are *h. pano* (very dark), *h. ʻula* (reddish), *h. uli* (dark blue), and *h. moelola* long stripe).

One, unnamed, is said to be whitish, with blue spots. Kepelino described the *moelola* as having a long black streak running along the top of the back from the snout to the end of the dorsal fin, a faded red streak along the edge of the tail fin, a clear yellow streak below the black streak, the ventral portion a clear yellow at the forward part of the body, fins pale. All *hilu* fish are gaily coloured and prominently striped lengthwise. Size varies from about 12 to 18 inches long.

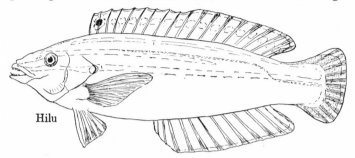

Hilu

Eaten raw, dried and salted, baked or broiled.

They are found in crevices of the reef, under large projecting *limu*-covered rocks, or asleep in the sandy bottom, completely hidden; they associate with *aʻawa* fish.

All Hawaiian informants remark upon the " quiet, ladylike " demeanour of the *hilu*. A child that is quiet from childhood up, is called a *hilu*, a pregnant woman who eats *hilu* will have a quiet, dignified child.

There is a legend of two *hilu* fish (75.44) :—

Two *hilu* fish were once supernatural beings, one named Kaululena, the other younger Maʻiʻo. The former had three forms, *hilu* fish, shark and man. Maʻiʻo had two forms, *hilu* fish and man. These two beings landed at a place near Kawaihoa, and there separated, Maʻiʻo going to the *koʻolau* (windward) side of Oahu, past Makapuʻu, and Kaululena going along to the *kona* (southern) side. Both travelled in their *hilu* forms.

Makaliʻi was the chief living at Hauʻula, on the *koʻolau* side, and had in his retinue several fishermen equipped with all necessary fishing gear—as was the custom. When Maʻiʻo swam along at Hauʻula he was observed by fishermen there and caught in their nets. He was dragged to the beach and the people shouted at the sight of so large a fish. The body of the *hilu* was cut into pieces and given to the people. Its blood ran into the ocean until the redness was reflected in the sky. The older brother, at Kaʻena, saw the redness spread over the sky and knew that his brother was dead. He resumed his human form and walked overland from Kaʻena. He came to a house where a piece of the *hilu* fish was being broiled. Without saying a word to the people he picked it up and threw it into the sea. And so he did at the next house where he found a piece of his brother's flesh, and the next and the next until he came to the house of an old man. The man was chewing *ʻawa*, and when the *ʻawa* was ready he began to pray in the name of the two *hilu* fish, Kauluena and Maʻiʻo. When Kaululena heard the prayer, he knew that this was their keeper (that is, a mortal guardian of a god—these were demi-gods). After the man had uttered the *amama* (the closing word of any prayer), he drank the cup of *ʻawa* (a ceremonial drink).

Kaululena asked the man, " Do you know these gods, Kaululena and Maʻiʻo, to whom you pray?"

" I have never seen them, but nevertheless I have imposed a *kapu* on myself, as I have been accustomed to do from the time of my parents and grandparents. This morning Makaliʻi's fishermen caught a huge fish, and all the people here except myself have gone home with the flesh of the fish. But I have not eaten of the flesh for I am the worshipper of the *hilu*, as you know by my prayer." He did not guess that the handsome stranger before him was Kaululena, one of his gods.

After the old man had finished speaking, Kaululena said to him, " Listen to the instructions that I am giving you. Put *lepa* (tabu flags) around your home and gather all your family within the space surrounded, for a terrible punishment is coming to this land, to those who ate the flesh of the *hilu*. I am the god whom you worship. The whole land will be flooded."

Kaululena stood up, left his keeper, and continued going about finding pieces of his brother's body. He went on till he came to the upland, to the stream of Kaipapaʻu. This is a circular spot, surrounded by hills, wide inside, and narrower downstream, the sides of the hills very close together. He prayed to the rains of the sky to fall directly upon the inclosure among the hills there, the opening of which he was damming up. And the rains fell. As the water rose in the hollow surrounded by the hills, the fish body of Kaululena grew in size, cutting off the water from flowing below. When the water rose until it almost reached the top of the basin, he leaped away and disappeared into a cave far up on the side of a hill, thus releasing the water, letting it tear down the stream bed and destroy the natives of the land. In the places where the flags had been set up, in accordance with Kaululena's command, that is, at the home of his keeper, the people within were all spared.

And Maʻiʻo came alive again in the form of the striped *hilu* fish which we see today. The streaks indicate where his body had been touched with fire, or slashed for salting.

Hinālea (occasionally shortened to *ālea*). Some of the fish called *hinālea* by Hawaiians have been identified as:—

Thalassoma ballieui (Vaillant and Sauvage), *h. luahine* (old woman) (greyish); *T. duperry* (Quoy and Gaimard), *h. lauwili* (*wiliwili* leaf) (yellowish), very common.

Coris gaimardi (Quoy and Gaimard), *h. ʻakilolo* (bite-brain or headache), which some have identified as the *Macropharyngodon geoffroyi* (Quoy and Gaimard), still others as the *Gomphosus varius* Lacépède, described at the end of the *hinālea* notes. One informant clears the difficulty somewhat by adding that there are two *h. ʻakilolo*, one greenish, one bluish.

Gomphosus varius Lacépède, *h. iʻiwi* (*iʻiwi* bird); *h. nukuiwi* (beak like *iʻiwi* bird); *h. nukuloa ʻeleʻele* (long black snout).

Other Hawaiian names for *hinālea*, not correlated with their scientific names are:—

Hinālea 'ele'ele (very dark, or black) ; *h. lipoa* (fragrant algae) (greenish) ; *h. mananalo* (bland) or *ananalo*, a very small variety; *h. ni'au* (this word has two meanings: coconut midrib, and spleen) ; *h. palau* (broad), this one has a broader body than the others, broad and flat.

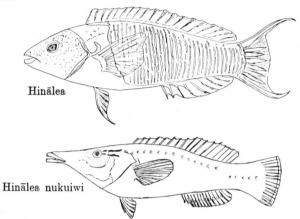

Hinālea

Hinālea nukuiwi

Hinālea are from 3 or 4 to 10 inches long. The colours vary greatly. *T. duperrey* is bright green with purplish-red, vertical bars, with a broad vertical band of reddish orange back of the head; head is bright blue, tail fin and pectoral fins blue, colours vary in specimens, upper and lower edges of tail longer than central portion. J. and E. (40, p. 295) say these are " beautiful fishes of the coral reefs and warm currents, the colouration largely deep green or blue." Scales are large, the head scaleless, skin slimy to the touch, difficult to rid of scales.

This is one of the most popular, abundant and well-known small fish. Hawaiians often mention it in tales and traditions as the proper fish to eat as an aftertaste for *'awa*. It was usually eaten raw. Because the scales were difficult to remove, the fish was usually skinned, before or after cooking. The favourite way of cooking it was broiling. It was also good for *i'a hoomelumelu*, in which case the entrails, head, tail and spine were removed, the flesh scraped off the scaly skin, the condiments added. A supply of *hinālea* could be kept at hand, for they live on sea plants and remain healthy if confined in sea pools. This use is mentioned in a story (75.14) : " When the *'awa* was ready to pour into the drinking cup, then someone ran with a round scoop net to a sea pool and scooped up some *hinālea, kūmū* and other fish that had been caught already to save the trouble."

Hinālea were used as offerings to the gods sometimes. For help in bringing on pregnancy, one is advised:—

> Take two *hinālea* fish, wrap in *ti* leaves and cook on the coals. The first fish is Ku's* and before eating it pray " O Ku, (mention the name of the hopeful woman) is going to eat the *hinālea* fish, an offering to you, O Ku. Grant us a child, an offspring of

* Note: Ku is one of the four most powerful gods of Hawaii.

yours in this world," and so on. If a son is wanted, mention the duties of a man, such as fishing, house building, etc., and likewise for a daughter, the duties of a tapa maker, etc. The second *hinālea* fish is Hina's (an important goddess) and the *kahuna* uses the same prayer. (75.8.)

A story of the origin of the *hinālea* is mentioned by Kamakau (47, Chap. 4, p. 55) in a tale of an *'e'epa* (in this case a goddess) woman who, angered by two supernatural beings who had helped her faithless husband, tore her enemies to pieces and the pieces became *hinālea* fish.

Hinālea 'akilolo: *Gomphosus varius* Lacépède.

Description: Size, 5-12 inches; colour, greenish, with brown tints toward tail, bluish underbody; fins greenish blue; colours somewhat variable. Mouth is small, a bait-teaser. One informant says some rather grey, with white spots on back, whitish below.

Eaten salted, dried, broiled over coals, or, best of all, wrapped in *ti* leaves and then baked or broiled; a delicious fish, flesh sweet.

This fish lives in shallows as well as in waters as deep as 7 or 8 fathoms.

The meaning of the name *'akilolo* (*'aki*, to nibble, *lolo*, brain) gave this fish value to the *kahuna*, who used it as a *pani* (final taste of food) in sicknesses of the head.

Hinana (many young), the spawn of *'o'opu*.

Hi'ukole (*hi'u*: tail; *kole*: raw, skin removed), see *'o'opu*.

Hi'u-'ula (red tail), see *'o'opu*.

Hoana (grindstone), see *kukala*.

Hou.   Two of the *Thalassoma* species are called *hou*.

According to Hawaiians, the main colouring is greenish, in both species, but one is striped with blue, the other with red. A Hawaiian description, by Kepelino (52), is " body green, with long, reddish stripes from front to back, colour on back a dirty reddish, the long stripes below are a clear, handsome red; dorsal and ventral fins covered with fine red stripes throughout; cheeks covered with red streaks and dots, or blotches." This description agrees with the colouring of cast No. 427 in Bishop Museum labelled " *Thalassoma purpureum* (Forskal)." The size is from 8 in. to 2 ft. long, usually about 15 inches; body plump. Scales are large and slippery.

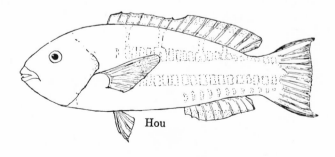

Hou

In the young stage, many names are given to the *hou*: *olani, olali, palaea, pakau-eloa, pakaiele, kanaloa, pa'ou'ou, pahouhou* (in Ka'u), *'awela, 'awala*. In Kane'ohe, Oahu, *olali* is the popular name. Kinney says that the young are of many hues; they dart about the pockets of the reefs.

The *hou* is classed as " delicious." Some say it was eaten raw, some say not. It was, however, cut into thick slices and cooked in *ti* leaves; also dried. The liver, mashed with *kukui* nut relish and chili peppers is reputed to be delicious as it slips down the throat.

The *hou* lives in shallow water, sleeps in sea pools at night at a depth of only about a foot or so, and snores like a human being. Its snoring is easily heard if it is approached quietly, and so it is fairly easy to catch it at night by spear and torchlight. Its habit of remaining in sea pools was used in its capture; nets were spread across the seaward opening of a pool, a noise made to scare the fish, and it was caught as it jumped to escape.

One informant says: " A peculiar fish—when a bad-tempered man saw it cutting capers while he was fishing, he went home to beat his wife." The behaviour of the fish was supposed to indicate the behaviour of the wife.

Huhune (small), see *'o'opu*.

Hukiki, see *puhiki'i*, small stage of *mālolo*.

Humuhumu (to fit pieces together), trigger fish: *Balistidae.*

A few Hawaiian and scientific names have been correlated:—

Melichthys vidua, h. hi'ukole (raw tail), or *h. uli* (bluish, dark), deep water.

Sufflamea bursa, h. umaumalei (lie on chest), said to have a bad odour; light in colour; small; deep and also shallow water.

Sufflamea capistratus, h. mimi (urine, malodorous), dark brown, about 8 in. long.

Melichthys niger, h. 'ele'ele (black), large; deep water.

Rhinecanthus aculeatus, h. nukunukuapua'a (nose like that of a pig).

Rhinecanthus rectangulus, also called *h. nukunukuapua'a.*

Other Hawaiian names are *h. me'eme'e* (small, black with blue on fin); shallow water, *h. kapa* (bordered), or *h. kapu* (prohibited) (small, dark); *h. mane'one'o* (irritating) (has a long horn; yellowish;

Humuhumunukunukuapua'a

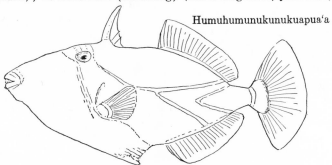

a deep water *humuhumu*). The *'ui'ui* (a small squeaky sound) is thought by some to be the young of the *h. mane'one'o*, but by others as different from it.

Description: *Humuhumu* vary from about 6 inches to 18 inches in length. Colour markings vary considerably. Many species are of more than one colour, the colours in blocks, marked off by stripes, or bands of still other colours. Some species are of one colour throughout. The colours of the fish sketched from cast in Bishop Museum, are light brown above, chest whitish, middle area posterior to chest dark brown, as also the area anterior to tail fin, intervening area light brown, the bands separating the blocks of colour are greenish yellow, base of pectoral fin is bright red, over the mouth and eye there are faintly bluish tinges. The skin is rough, hard, tough, with rather large rough scales or scutes; the body much compressed. The *'ui'ui* is described as a small fish, 2-3 inches long, having scales "softer than the *humuhumu*" (Pukui), grey-green in colour, with yellow spots (Leong). Another informant (Watson) says it was yellow, dotted with black, which is the colouring of the *'oili lepa* (raised banner). Three informants agree, Kaauwana Aukai, William

Watson and Beckley (4, p. 6) that it " makes its appearance at intervals of from ten, fifteen to twenty years." It was good to eat when it first came in to shore. Then it was fat, with liver much enlarged. After a month the fish become thinner and then tasted strong and rank. Some never ate it—" Tza! firewood "—but used it for fuel to cook better food. The coarse skin was pulled off, the flesh salted, and sometimes *manauea limu* (sea weed) added. They were caught by lowering a basket with cooked pumpkins or sweet potatoes as bait into a school. They ate greedily.

All *humuhumu* have a strong smell, like a pig, and must be skinned before eating. Broiling in *ti* leaves is the favourite method of cooking and in modern times by frying. Kepelino (52) says the flesh is good, better than the *manini*; another says it is too bony to be considered a good food fish. Japanese in Hawaii are fond of it, and it therefore brings a good price in the market. Some *humuhumu* grunt like a pig when pulled out of the water. Hawaiians used *humuhumu* for fuel when they were caught in great abundance. A few cooked, eaten, then the bones, especially of the head, were used to keep the fire going for further cooking. Though some *Balistidae* are violently and dangerously poisonous in other parts of the Pacific, they are not poisonous in Hawaii. (83.) The *h. mane'one'o*, however, causes a puckery feeling in the throat—one of the marks of a poisonous fish.

I'a holokola, see *kunehi.*

Iaku, listed (59 a).

I'a-makika (mosquito fish) : *Gambusia affinis* (Baird and Girard), introduced.

In 1905, the City of Honolulu asked Dr. D. S. Jordan to search for a fish that would consume mosquito larvae. He sent Alvin Seale to Texas to get fish of three species. One of these, *Gambusia,* proved the best enemy of mosquitos, and since then has been widely used. (53, pp. 84-85.) The fish became plentiful enough to supply Hawaiians with another fish to eat, though a small one. The fish were salted and dried or cooked in *ti* leaves.

'Iāo (*'i'iāo,* or *'iomo*), silversides: *Hepsetia insularum* (Jordan and Evermann).

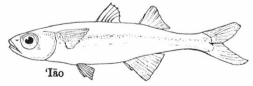
'Iāo

Description: Size, up to 2, sometimes 3 inches; colour, bluish white; scales large.

Eaten in some localities, even esteemed. Usually eaten raw, sometimes salted and dried. Scales were removed by putting many of the fish in a vessel with coarse sand or small pebbles, stirring them around until the scales came off, then rinsing them in sea water. Not eaten in Kane'ohe Bay region.

The *'iāo* is abundant in shallow pools and bays inside the reefs, and is used chiefly for bait for such fish as *aku*. It is phosphorescent. The face of a human victim for the *heiau* was sometimes rubbed with tiny *'iao* so that it shone like the eyes of the maneater shark of the deep.

I'a-pākē (China fish): *Ophiocephalus striatus* Bloch, introduced.

Not eaten by Hawaiians.

I'a-'ula'ula (red fish) : *Carassius auratus* (Linné), the goldfish.

In the time of Ka'ahumanu, goldfish were introduced from China (possibly about 1810). Ka'ahumanu kept them in a pond in Manoa Valley where she had a home, placing a tabu on them until they became established. It became the common custom to breed them in taro patches and in the watering troughs or barrels for stock animals. They were distributed to all the islands. Hawaiians used to eat them with relish. When they were fully grown, about 8 inches, they were prepared by scaling and rubbing the skin with salt to rid it of the unpleasant, muddy, fresh-water taste, then cut up for eating raw with " lomied " raw onion and salt. In modern times such fish were also fried (Lizzie Maka). The goldfish with white markings is the *i'a-'ula'ula lākea* (white dorsal); the *i'a-'ula'ula uli* (dark red fish) is the carp, *Cyprinus carpio*, introduced, well liked by Orientals.

Iheihe (spear-like), the half-beak. Three species.

Euleptorhamphus longirostris (Cuvier).

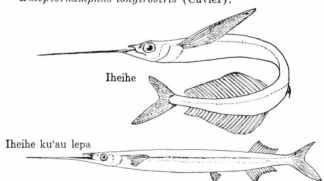

Iheihe

Iheihe ku'au lepa

Jordan and Evermann (40, p. 128-130) give its size as 16-18 inches; colour, pale, bluish, silvery above; scales on back with darker edges, lower side of belly silvery, top of head dark bluish, "bill" bluish, fins pale bluish, anal white; head compressed, flattened on top and the lower surface narrowly constricted. Thompson labels Bishop Museum cast No. 102, "rare, pelagic." Another *iheihe* is the *Hemirhamphus pacificus* Steindachner. This may also be called the *a'u kuau lepa*. Its size is 8 to 10 inches long; colour, lower part of body silvery, darker above; lateral line marked by a band of blue; scales rather large. The third *iheihe is H. brasiliensis* (Linné), also called *me'eme'e*. It is 13 to 15 inches long, the colour is brownish on back, shading to pale green at the lateral line, which is marked by a deep blue streak; underbody silvery; beak very dark. (From cast No. 103 in Bishop Museum). In Bishop Museum casts, the "bill" of 102 is very long, that of No. 103 is short. One other Hawaiian name belongs to this group, *pu'ili* (grasp together). It is said to be more roundish than the true *iheihe*, which is flat-bellied and round backed.

Not eaten raw in Ka'u; favourite way of preparing is broiling in *ti* leaves.

The young of these fishes sleep on the water at night, within the reefs, and gleam in the light of a torch like a pencil of brilliant, metallic blue.

'I'i (tiny), see *ma'i'i*.

'I'iao, see *'iao*.

'Ilae, may be another name for *'ulae*.

'Iomo, see *'iao*.

Ka'ape'apehā, or *ka'ape'ape*.

This name appears on a list of fishes in a Hawaiian newspaper (59 h), and is also described by Kepelino (52):—

The general appearance of this fish is like the *moano*, but the *moano* is red, banded with black, and this is entirely a dark red. This is a deep sea fish. It is tasty and delicious to eat. There are two kinds, the red and the black.

Kahala (*amuka* may be another name) the amber fish: *Caranx mate* Cuvier (*omaka*); *Seriola aurea-vittata* Schlegel, *kahala opio* (young), *S. dumerili* (Risso), and *Naucrates ductor* (Linné). Other Hawaiian names are *kahala mokulei*, or *mokuleia* and *kahala maoli* (indigenous). Young stage, *puakahala* (*amuka* may be a term for this stage only), half-grown are *k. opio* in the opinion of some, adult, *kahala*.

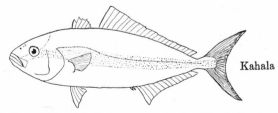

Kahala

Description: Size, 6 ft. or more in length, body is slender; scales small; colour is purplish blue which reflects light like silver; a yellow stripe runs through each side from pectoral fin to tail.

Flesh is white, fine and soft. It is cooked whole in the *imu*, or in steaks, or salted, or wrapped in *ti* leaves and baked. In some localities eaten raw, first skinning the fish, cutting into cubes, then salting—a meaty fish.

Kahala are deep sea fish. Many are caught off Niihau, sometimes 40 hooks being let down on one line. The following proverb reflects esteem: *Pololei a'e la no a ka waha o kahala* (It went straight to the mouth of the *kahala*; that is, it went straight to the high chief). (46, No. 119.)

Kahala manini, see *manini.*

Kakaho'oulu (drawn from the depths), listed 59 h, 8/26/99.

Kakaia (fished from the deep sea), listed (59 g).

Kākū (another name is *kuapala* or *kupala,* but used rarely, used in Molokai; common name is *kākū*), the barracuda: *Sphyraena barracuda* (Walbaum). The meaning of *kākū* is to prod, see also under the *a'u* fishes.

Kākū

Description: 4 to 6 feet long; colour, J. and E. (40, p. 142) say dark olive-brown above, sides silvery, about 20 very faint blackish bars just above lateral line. Cast in Bishop Museum is nearer grey blue in colour, lighter underneath, fins all dark blue, and a few dark spots scattered casually below lateral line.

According to various reports it was sometimes eaten; one says highly relished, another says eaten broiled but not eaten raw; flesh is fine, white. Chinese in Hawaii are fond of it.

The fish is common in some areas, rare in others; one informant says deep sea only, seldom seen near shore, another says that it sometimes gets into fish ponds, is difficult to get rid of and is vicious toward other fish. In ponds they get fat, weigh 20 to 30 pounds at 3 ft. in length. They are strong enough to pierce through a 1½ inch wire mesh. In Puna they are reported to have bumped against canoes sometimes, usually at night when a canoe has a light in it.

Kala (rough); (names less well known are *ume* (draw, pull) and *mahaha* (broad)); surgeon fish: *Acanthuridae* species: *Naso lituratus* (Schneider), *N. unicornus* (Forskal), and *N. brevirostris* (Valenciennes). Hawaiian names of varieties are: *kala holo-ihu-loa* (long-nosed); *k. li'ili'i* (little), also called *k. pahikaua* (sword); *k. lolo* (imperfect) (*N. brevirostris*); *k. moe* (sleeping); *k. moku* (cutting), also called *k. awa pehu* (swollen); *k. ni'au* (spleen); *k. palaholo* (spry); *k. ilu* (insignificant); *k. holo* (swift).

Today, in Kane'ohe Bay, there are three *kala,* the large ones called merely *kala,* 7 to 8 pounds, and 14 to 15 inches in length, brownish, with a horn; the young of these are called *pakalakala* (partly grown *kala*), they have a yellow tail when they are 4 to 5 inches long; the second is *kala oheno* (a little larger), when small they are black, when full grown they are light grey, they are about 14 inches long and weigh 6 to 7 pounds; the third (special name lost) is small, yellow and never grows longer than 4 to 5 inches. The young are called *pakala, pakalaka,* or *pakalakala* at hand length and may weigh ½ to 1½ pounds.* In this stage they are broiled over coals.

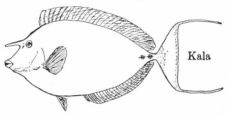

Kala

Description: 12 to 24 inches long, body much compressed; colour varies, some as given above, some yellowish grey, fins touched with light blue and reddish brown; the *kala lolo* is black. The skin is without scales and is tough—tough enough to have been used for the tops of small, coconut shell knee drums. All but the *kala lolo* have a single, horn-like spine on the forehead, which develops as the fish reaches maturity. In the *kala lolo* it scarcely projects beyond the snout. The knife-like projections just before the caudal fin are very sharp and can wound painfully.

The *kala* is a very popular fish with Hawaiians despite the fact that, or possibly because of the fact that the flesh has a strong odour. The meat is white. Few eat it raw. If cooked fresh the favourite method is broiling. The soft parts make good *palu*. It is also dried, the tough skin first stripped off (*uhole*), the flesh then cut off in " ropes " and dried, or the meat cut away from the spine and dried, skin and all. If the skin is left on the flesh is cut down to the skin, after removal from bones, so as to let the salt sink well into the flesh. Kukea says it is best broiled or baked after it is about three-quarters dried, that is, not too stiff, and hard. This fish is so abundant that it is eaten considerably for that reason. Mullet and *moi* and other fish are more popular, but *kala* is easier to find and catch.

The odour of the flesh is said to vary with the area where it is caught. Those from the region of Mokoli'i islet, Kane'ohe Bay, are strong in odour, but from Wailau to Mokapu, in that bay, the *kala* are almost without disagreeable odour. To rid the flesh of odour, the fish was treated as were the *palani* and *pawalu*. The directions are:—

Lay the fish across the palms of the hands, head on the left palm, tail on the right. Breathe (inhale) over the fish, turning the head from left to right, and then expel the breath violently. Turn the fish over, and repeat. (Lily Akana.)

In localities where *limu* (seaweed) is especially abundant, such as Keahi, Kailua, and Waikiki, Oahu, the *kala* takes on the fragrance of the *limu* that it eats. Where the *limu kala* was abundant, and the

* Hand length was one unit of measure for Hawaiians, about 5 inches.

fish too, sweet potatoes were often let down into the feeding grounds for a week or two, then the fish might be taken, some at a time, without disturbing or alarming the rest of the fish.

In a legend of ghosts (75.47), the hero, Punia, rolled up some ghosts in a fish net, and killed them. The killing of the ghosts was said to be the cause of the occasional phosphorescent glow (*weli*) on the water and the strong odour of the *kala* and *palani*. In the story of Lonoikamakahiki, the fish is mentioned: " The *kala* shall be my fish . . . It is a fish sacred to my god." (25, Vol. 2, p. 288.)

Kalaloa (long spike), a species of the *'ala'ihi*.

Kale, see *kole.*

Kalekale, see *opakapaka.*

Kalikai, see *ukikiki.*

Kamano (salmon).

This name was applied to the introduced rainbow trout (streams of Kauai), to the introduced salmon, which did not succeed in establishing itself, and to the salt salmon imported in large quantities from the Northwest Coast of America. According to Mrs. Lahilahi Webb, the Hawaiians who shipped as sailors in the days soon after discovery (1778), often got to the Northwest Coast. They took a fancy to the salmon there, and salted and brought back some. It became popular at once, and soon became one of the foods kept on hand in Hawaiian households. For the early missionaries it took the place of salt cod, dear to New Englanders. It was prepared by soaking in fresh water to remove some of the salt, then following New England recipes or native ways. A piece of salt salmon is often included in a leaf-wrapped bundle of pork and greens (then baked in the *imu* with other foods). One popular dish developed is salmon-lomi. (Preparation of lomied fish is described on p. 21.) For salmon, raw tomatoes were added, and raw onion is a necessity. In modern times chipped ice is often added, though to my taste this is a mistake as it melts and thins the flavour. For many years no Hawaiian feast has been complete without salmon-lomi among the side dishes.

Salt salmon became the mainstay of the poor people in the old days, when other fish became scarce. It is a hardship to Hawaiians when the supply is scarce, such as during the war, 1941-1945, and when the price is high.

Kamanu, called the sea salmon in Hawaii; one of the amber fishes: *Elagatis bipinnulatus* (Quoy and Gaimard).

Description: Size, a cast (No. 19) in Bishop Museum is about

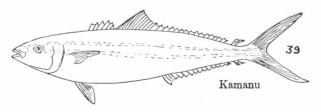

39

Kamanu

31 inches long; J. and E. (40, p. 186, no. 138) say their description is based on one specimen, 3 feet long, obtained in the market (J. and E. were in Hawaii from June to September); Fowler (26, p. 143) examined 46 specimens in the market in November and December, the largest about 3 feet long. Colour notes from cast 19 are: deep sea-green above lateral line, fins lighter in colour, ventral, anal and pectoral whitish, a single purplish-blue streak from head to tail, lower than the lateral line, colour between lateral line and this streak is light green, below the streak it is yellowish white; eye is yellow. A coloured plate (27, No. 61, fig. 3) in Quoy and Gaimard's atlas, of a fish about 9 inches long is greenish-blue above, almost white ventrally, fins all bluish green, with touches of yellow along body above anal fin, as well as at base of pectoral and the eye; two vivid streaks from head to tail. J. and E. (40, p. 186) speak of two streaks. Scales small and numerous.

No Hawaiian notes about this fish; probably used similarly to the *kahala*, another amber fish.

Kanaloa, see *hou*.

Kani'o (striped vertically), see *'o'opu*.

Kapā, see under *puhi*.

Kapuhili (confusing), see *kikakapu*.

Kauleloa (the long penis), see *'o'opu*.

Kaumakanui (big eyes placed on it), the sunfish: *Mola mola* (Linné). Not eaten by Hawaiians; Orientals appreciate it.

Kawakawa, bonito, little tunny: *Euthynnus alletteratus* (Rafinesque). By Hawaiians sometimes called *poho-poho* (patches) because of black dots, size of fingernail, on belly. The young called *kina'u* (immature); second stage, *'ahua* (larger); adult stage *kawakawa*. The *aku* shares these terms, *kina'u* and *'ahua*, until adult, for the two fish are difficult to tell apart until maturity.

Kawakawa

Description: 3 to 4 feet in length; body robust; colour, in general a cold blue.

Eaten raw or cooked in any way; flesh is red; highly esteemed. Of the bonitos, this is the one liked best for eating raw.

Kawēleā (long and bright) (or *halalua*): *Sphyraena helleri* Cuvier. A small relation of the *kākū*.

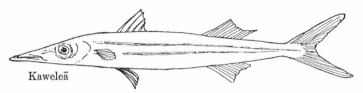

Kawaleā

Description: Length, 16 to 26 inches; colour, blue, deep blue on back down to first lateral stripe, the rest of the body pale, silvery blue, or bluish grey, except for tail fin and upper part of head, which are also deep blue; the two lateral stripes are a soft yellow.

This is a particularly delicious fish, with very few small bones. The flesh is white. It is good broiled or cooked in any other way; eaten raw when it is fat. The flesh is softer than in the *kākū*.

Like the *kākū*, the *kawaleā* is destructive to mullet ponds (40: 143). Notes differ as to abundance, one says rarely found near shore, another that it is common in the reefs. It is usually caught at night.

Keiki o ka manō (child of the shark), see *leleiona*.

Kēkē (big bellied), see *'o'opuhue*.

Kiahamano, see *'o'opu*.

Kihikihi (angular): *Zanclus canescens* (Linné) ; *Chaetodon auriga* Forskal; *Zebrasoma veliferum* (Bloch). Hawaiian varieties are *kihikihi launui* (big-leafed), or *mane'one'o* (irritating), *k. alo-'ula* (red breast), silvery, *k. pohaka* (big spot) and *k. halena* (yellowish).

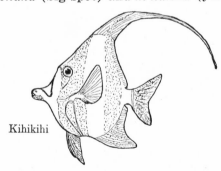

Kihikihi

Description: Length, about 3 to 4 inches, some as long as 6 inches; colour brilliant, black bands alternating with yellow, edged with blue, snout has a touch of orange. Skin is rough, without scales.

One source says not eaten much, too little flesh, " Bah! it is a *kihikihi*, a poor-fleshed fish, and a waste of time to broil." (Pamano story.)   Another says, " A delicious fish when broiled, not only to the back-country people; it is considered delicious in the courts of the chiefs, delicious to Panaha'eke (Panaha-the-humble)" (52).   The meaning is that this fish is delicious if better may not be had.

Kikakapu (*kika*—energetic; *kapu*—prohibited): *Chaetodon ornatissimus* Cuvier, *C. fremblii* Bennett, *C. lineolatus* Cuvier (these three called *kikakapu kapuhili* (confusing) by some), *C. lunala* (Lacépède), *C. unimaculatus* Bloch, and *Cheilodactylus vittatus* Garrett, which may be the *k. koʻa* (coral). There is also a Hawaiian name unattached to a scientific name, the *k. nukuʻoiʻoi* (pointed beak).

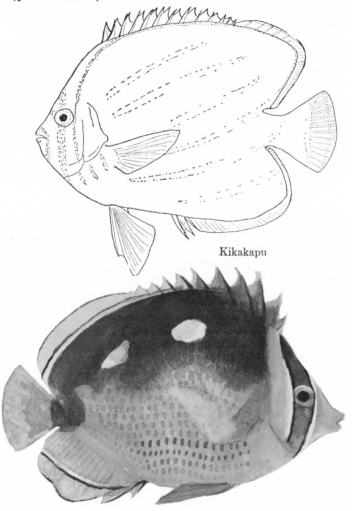

Kikakapu

Description: Length, about 6 inches, many only 3 to 4 inches; gaily coloured, the *C. ornatissimus* is creamy all over, with bright yellow head and fins, touched with black; body streaks are deep orange, as is pectoral fin. Others vary considerably, one being brownish, with reddish eye, some are yellow, streaked with blue, with a large black blotch near tail.

Not eaten much because they have so little flesh. One allusion to this fish is, " Marks of the *kikakapu*, the sacred fish with the bitter gall." (25, I: 240.)

Kina'u, this term used for the spawn of both *aku* and *kawa-kawa*, as well as for a small eel, and one of the *'o'opu*.

Koā'e (tropic bird), see under *weke*. When large, this variety of *weke* is sometimes referred to as *koā'e*, rather than *weke koā'e*.

Kohola (hump dorsal), humpback whale, not used for food; carcass called *palaoa*.

Kapu to common people, reserved for chiefs, who valued the ivory teeth and the bones. If a *palaoa* drifted ashore or grounded on the land of a chief it became his property. The term *palaoa* also denoted the ornament made of whale's tooth. Green (28, p. 15) records that Hawaiians believed that a whale leaping and blowing presages a storm. (Included because it is *i'a* though not a fish.)

Ko'i (adz), see *ulaula*.

Kokala, see *kukala*.

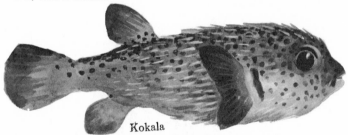

Kokala

Kole (raw, skinned) (sometimes *ukole*; the *kale* of J. and E. (40, p. 398) is probably incorrect): *Ctenochaetus strigosus* (Bennett). Referred to poetically as *kole maka onaona* (bright-eyed *kole*).

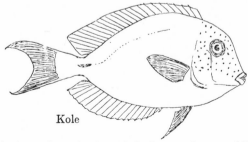

Kole

Description: Length, 4 to 6 inches; colour, reddish black; a sharp spine, depressible in a groove, is just in front of the tail; scales very small.

One informant says they were never cooked, but were eaten raw, relished for their flavour.

A tradition of house building is that the *kole* was put in the ground where the houseposts were to stand, that is, the posts on the side facing the east. Should a *kahuna* enter and predict trouble for the householders, he would die.

The *kole* run in schools, sometimes with the *pakuikui;* they are abundant in Kona.

Kolekolea, mentioned by Kamakau, nothing known about it.

Kolokolopā (like an advancing wall), the young of the *kūmū* and the *weke,* and any other goatfish.

Kowā'e, another spelling for *ko'āe,* see under *weke.*

Kowali (morning glory), a kind of *puhi* (eel).

Kuapala (yellowish back), see *kākū.*

Kukala (*kokala*) (spiny) or *hoana* (grindstone), porcupine fish: *Diodontidae—Diodon hystrix* Linné, *Diodon holo-canthus* Linné and *Cheilomycterus affinis* (Günther).

Description: Length, about 9 to 21 inches; body is covered with spines, which are like long quills sticking out like those of a porcupine when the body is inflated. Spines capable of inflicting a painful wound, and regarded by some Hawaiians as actually poisonous. The flesh is not poisonous, according to one informant, others say it may be as poisonous as that of the other puffer fishes, another says the *hoana* is the "smooth-skinned Diodon, and is deadly poisonous." All the species listed by Jordan and Evermann are spiny, none have smooth skin.

One form of a sea god was named *Kane ko kala.* Those who had this god as *'aumakua* threw the fish back into the sea if they caught it, with the words:—

E Kane-kokala	O Kane-kokala
E Kane-kokala	O Kane-kokala
E ho'i, e ho'i	Return, return
I Kane huna-moku.	To Kane-huna-moku.

Jordan and Evermann (40, p. 438) record the name *'o'opu kawa* for one of the Diodons.

Kule, see *'ulae.*

Kūmū, a goatfish: *Parupeneus porphyreus* (Jenkins). Stages of growth: *kolokolopao,* or *kolokolopā* (like an advancing wall), or *makolokolopao,* the spawn stage. One informant says that *kolokolopao* is the name of a fish "between the *weke* and the *kūmū,*" a little darker red, has no white on tail, has a smaller mouth; the *kūmū* is lighter in colour, with a white streak on tail, and the *weke* is red all over. The second stage is *ahuluhulu,* the third *kūmū-a'e* (next to *kūmū*), and the adult stage *kūmū.*

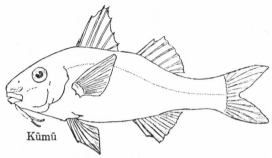

Kūmū

Description: Length is from 5 to 16 inches, usually from 9 to 12 inches; colour, red, the back a bright, rosy red, the underpart a little lighter. Those caught in the deep sea said to be of a deeper colour than the shallow water fish. Scales large.

A delicious, abundant, popular fish, highly esteemed. It was sometimes kept in pools to have on hand to eat after drinking *'awa.* Eaten broiled, cooked in *ti* leaves, raw, or salted lightly for two or three days, then cooked in any manner.

The *kūmū* was used extensively as an offering to the gods when the priests demanded a red fish. It was an appropriate offering when a canoe was launched, sometimes in hula ceremonies, and sometimes for atonement of sin. It was offered by those who had been through a course of teaching and were now " master " of an art, for one meaning of the name is *master.* It was classed as one of the " sea pigs " (see p. 43 herein). It was forbidden to women for the red colour suggested the menstrual period, at which time women were set apart. The young, *ahuluhulu,* were used in the rite called *pa'ina ho'oku'u* (remembrance of the *'aumakua*) when the priest had delivered an afflicted person from death.

Kuna; the ancient name for eel.

Kunehi (apahu (cut off), *makua,* or *i'a holokola*) ocean sunfish: *Ranzania truncata* (Gmelin), *Mola mola* (Linné), *Masturus lanceolatus* (Lienard).

Description: Very large, may weigh as much as a thousand pounds; Grey (29) says: " Grows to a length of ten feet, a height of twelve feet, and a weight of more than a ton." The young, about a foot long, have been noted in reef waters in late summer, then they go out into deep waters, says a Hawaiian informant. J. and E. say (40, 439)· " Fishes of the open seas, apparently composed of a huge head to which small fins are attached." *Ranzania* is described as bright silvery on sides, upper parts dark, sides with brighter silvery bands " with black spots and borders. (J. and E. 40: 440.) This fish is exquisitely beautiful in colour, the colour changes are swift as the fish dies. " Body oblong, covered with a rather rough skin." (26, p. 475.)

Probably not eaten to any extent, partly because it is fairly rare. Some Hawaiians say not eaten, one says it was eaten. Jordan and

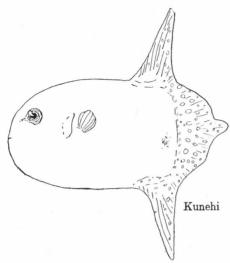

Kunehi

Evermann say (40, p. 439) " The flesh of these fishes is coarse and tough and not used for food."

Kunounou, or *kupoupou* (chunky), see *kupou.*

Kupala, see *kākū.*

Kupīpī (*oʻonui,* or *aoao-nui* (big sided)) ; some say the *oʻonui* is more like *the hinālea*) : *Abudefduf sordidus* Forskal, and probably other *Abudefduf. Aoao-nui* may be the name of the young.

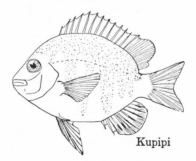

Kupipi

Description: Size, palm of the hand; body much compressed; colour, grey, distinguished by a black spot on the dorsal part of the caudal peduncle; colour graduates to lighter below the wavy line indicated on figure.

Meat is white, delicious, good to eat raw, cooked in *ti* leaves, or broiled, also dried. One says preferred raw after being salted for two hours or more; or broiled after being exposed to the hot sun for a day. Skin is tough.

A common fish in the coral reefs; plentiful, easy to catch;

perhaps eaten for that reason. One informant says, " Fisherman try
to evade hooking and bringing it out of the water, let alone packing a
load home."

Kupou, kupoupou, or kunounou, or poupou (short and
plump) : Cheilio inermis (Forskal). One variety called
kupoupou-lelo (yellow). Spawn are called ʻōhua (along
with spawn of many other fish).

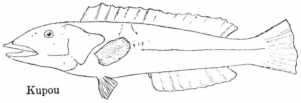

Kupou

Description: About 1 ft. in length, body narrowly compressed;
colour, olive-green, varying toward rusty red, scales touched with blue,
head with various streaks of these colours. Subject to great variation in colour. (40, p. 314-315.)

This is a good food fish, eaten raw or cooked in ti leaves.

Kupou are found in quiet waters, mossy places, for they eat seaweed and small crabs within the weed bunches. They may sleep on
the sand at night.

Common in the markets of Hilo and Honolulu. (40, p. 135.)

Laenihi (narrow forehead) : Hemipteronotus melanopus
(Bleeker), H. pentadactylus (Linné), Iniistius pavo,
I. aneitensis (Gunther), I. niger (Steindachner) and
perhaps others. Hawaiian names are laenihi ʻeleʻele,
colour very dark brown throughout, probably I. niger;
1. nēnē, scales marked like a Hawaiian goose, which is
speckled; 1. kea, whitish, perhaps H. pentadactylus,
which is yellowish white; 1. pukea, greyish, perhaps
the young of H. pentadactylus, which is greyish,
possibly the 1. pavo, which is greyish blue.

Description: Length, 6 to 14 inches, body compressed. As in
figure, the first two dorsal spines are set forward, almost disconnected
from the rest; they may, however, be reduced to a lower height than

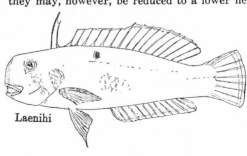

Laenihi

the rest. Colour varies considerably, as indicated above; characteristic of most of them is a dark blotch on the dorsal edge of body, midway, or some large blotches in the middle part of the dorsal fin. Scales large and thin, head almost without scales.

Good to eat, flesh very white, cooked in any way, evidently not eaten raw.

Laenihi live at about 25 fathoms depth, over a sandy bottom. This fish is honoured by being a character in the tale of Halemano. (25, II: 228-272.)

Lai (lae): *Scomberoides sancti-petri* (Cuvier). Young called *palaelae*, or *palailai*.

Description: Length, 3-16 inches; colour, silvery, bluish above, whitish underneath. No scales. The skin sometimes used for the tops of small coconut drums.

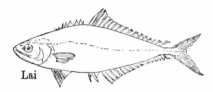

Lai

Delicious fish, broiled, dried, or baked in the *imu*.

A shallow water fish that travels on the surface. J. and E. report that it is not common, but *S. toloo parah* is fairly common, and that there is little difference between them. (40, p. 180-181.)

La-'i-pala (yellowed *ti* leaf) (*lau'ipala,* or *lau-ki-pala*): *Zebrasoma flavescens* (Bennett).

Description: Length, 3 to 7 inches, body much compressed; colour, a bright, lemon yellow, a beautiful fish; caudal spine strong; no scales, skin both delicate and rough.

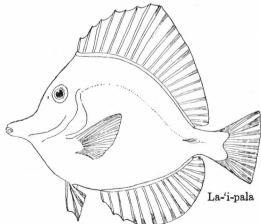

La-'i-pala

Usually eaten broiled; Kepelino says the skin is edible if cooked, but when eaten raw the skin is stripped off and thrown away.

Common about many reefs, always noticeable in Kealakekua Bay, a throng of them is a beautiful sight.

Lala, a small bait fish (25, vol. I: 378).

Lālākea (white fin), see *manō lālākea*.

Lāʻō (sugar cane leaf) (*ʻohua paʻawela* (like the *awela*)): *Halichoeres ornatissimus* (Garrett) and perhaps other *Halichoeres*.

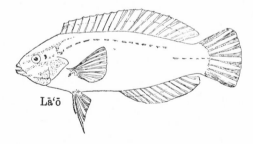

Lāʻō

Description: From cast 455 in Bishop Museum, length, about 5 inches; colour, general impression is green overlaid with dark, brick red at edges of large scales; dorsals and anal fins dark purplish red with long green band near outer edges, also at base of the dorsal; caudal the same but the streaks are more mottled, ventral fin green, pectoral transparent, head green streaked with orange red, a touch of crimson at base of pectoral. J. and E. (40, p. 285-287) say the species are numerous, of rather small size, and gay colouration, body slender, greatly compressed, colours even more varied than in Bishop Museum cast No. 455; sizes mentioned are 5 to 6 inches, apparently not very common.

Lapa (incessantly active), see *mahimahi.*

Lauhau (*hau* leaf): *Chaetodon quadrimaculatus* (Gray).

This and some other species of *Chaetodon* are known as *lauhau*, with these following descriptive secondary names: *lauhau kapuhili* (confusing), *l. kikakapu* (probably the same as *kikakapu*), *l. mahauli* (dark gilled), *l. wiliwili* (*wiliwili* tree), very dark with purple cheeks, *l. nukuiwi* (pointed nose).

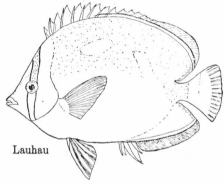

Lauhau

Description: Length, 3 to 5 inches, body much compressed; colour, *C. quadrimaculatus* is yellow, with upper half of body grey, the two blotches outlined in the figure almost white, dorsal and ventral fins with a band of pale blue inside the black line. The band through the eye is orange, edged with black and also blue narrow bands. Caudal, ventral and pectoral fins yellow. All species of *Chaetodon* are gaily coloured, yellow, orange, pale blue, grey predominating, with touches of black or dark grey.

Food value of *Chaetodon* is a matter of disagreement, one says, " bony, not worth eating "; another says, " sweet flesh, broiled on charcoal immediately, without scaling or cleaning." This is one of the fish used for *ho'omelumelu.*

Fishermen do not like this fish, for it is a bait nibbler. It is found in both deep and shallow waters, where it takes refuge in reef cavities.

Lauhua, see *o'opu hue.*

Lauia, this name recorded by Jordan and Evermann (40: 355-56) as *Callyodon lauia* (Jordan and Evermann). No data have been received from Hawaiians as to this fish. It may be the *uhu lauli*; see under *uhu.*

Lauʻi-pala, see *laʻi-pala.*

Lau wiliwili (*wiliwili* leaf) (*lauhau wiliwili*) : *Chaetodon miliaris* Quoy and Gamard.

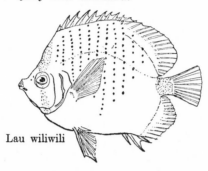

Lau wiliwili

Description: Length, 1 to 6 inches; colour, pale yellow, with a black blotch or band across the caudal peduncle, and another band of black from dorsal to ventral sides of the head, through the eye. Rows of pale blue spots string downward from dorsal to ventral part of body. Meaning of the name is " leaf of the *wiliwili* tree," an indigenous tree with flowers of brickish red, or orange or yellow. Scales are of moderate size, and merge into the dorsal and anal fins. The mouth is retroussé and the upper lip has the air of being turned back.

Eaten when better cannot be had.

Laukahiuʻu, see *manō laukahiʻu* (broad tail).

Laukipala, see *laʻi-pala.*

Lauli (dark leaf), one of the *ulua.*

Laumilo (*milo* leaf), a kind of eel, see *puhi laumilo.*

Lehe (large lipped), a deep sea fish, resembling the *ulua,* but fins are longer, skin darker. Eating and preparation the same as for *ulua.*

Leihala (*hala* wreath), see *puhi leihala.*

Leleiona (wild, behaving to attract attention), (*hehena* (crazy), or *keiki o ka manō,* child of the shark), shark-sucker, remora: *Echeneidae* species. At least one species present in Hawaii. Not eaten. It clings to large fish, such as swordfish, marlins, bonitos and sharks. Some Hawaiians thought it was the young of the shark.

Lelepo (far from land), see *mālolo.*

Lolo (brain), see *hinālea lolo,* or *ʻakilolo.*

Lolo-hau, probably intended for *lolo-oau.*

Lolo-oau (*pinao*) (dragonfly), flying gurnard: *Dactylopterus orientalis* (Cuvier).

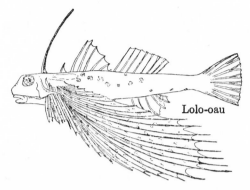

Lolo-oau

Description: Length, about 6 inches, sometimes as long as 14 inches; colour, bluish or drab above, white below, large pectoral fin spotted with " dusky golden spots " (40: 474).

Probably eaten, but no record found. Its flight is " curved " and it " drags its tail; it is found with the young of the *mālolo*." (Nakuina.)

Louia, another spelling of *lauia*? Another fish?

Loulu (name of a palm, name may have been given because of a colour resemblance—greenish white): *Alutera monoceros* (Osbeck), *A. scripta* (Osbeck).

Loulu

Description: Length, according to Fowler (26, p.461), *A. monoceros* reaches 360 mm. The colour in alcohol is brown, mottled with darker brown. (40, p. 423.) A cast of *A. scripta* in Bishop Museum is about 27 inches long; colour, greyish green, with irregular, medium dark blue spots, half circles over all the body, streaks on head; dorsal and ventral fins are a dirty greenish white, pectoral fin tan, the lone dorsal the same colour as the body, grey-green.

This fish was used by *kahuna* to cause death; it was referred to as " *he i'a 'awa'awa* " (a bitter fish).

It comes abundantly but erratically as to season.

Luhia, see *manō*.

Lupe (kite), discussed under *hīhīmanu*.

Mahae (divided), according to the Hawaiian dictionary, " a species of . . . fish of the *lauipala* group."

Mahaha, see *kala*.

Mahao'o (strong brow). No data.

Mahihi, see *mahimahi.*

Mahimahi, dolphin: *Coryphaena hippurus* Linné.

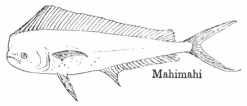

Mahimahi

Description, from cast in Bishop Museum: Length, a little over 5 ft.; colour, sea blue-green above, including the high dorsal fin; ventral portion yellowish; ventral and anal fins green, with yellow spines, caudal fin greenish; head silvery in front, yellowish underneath, the cheek green; pectoral green and yellow; about 30 dark blue spots on lower half of body, especially numerous and prominent below the pectoral fin. The head of the male *mahimahi* is vertical at front edge, the Hawaiian expression for this being *mahimahi lapa,* "like the prow of the ship *Claudine*"; that of the female slopes backward at the front edge, which the Hawaiians call *mahimahi oma,* "like the prow of the ship *Maunaloa.*" Both of these vessels were inter-island sailing vessels of long ago, and familiar sights to Hawaiians of their day. This fish is a beautiful sight as it goes flashing through the water just under the surface, at express train speed. Field says (24, p. 90): "The dolphin is remarkable for its brilliant and changeable colours: the colour of the fish in life is a dazzling silver, with yellow, green, and brown spots on the lower parts. After death, only faint indications of the former colourings remain."

Not eaten raw. Cut into slices and broiled over coals, also dried then cooked. This is the one fish for which there is a note that the "roe is delicious, either raw or dried, then cooked." Foreigners are fond of *mahimahi* steaks, usually broiled.

The *mahimahi* is a voracious deep sea fish that haunts the *'opelu* feeding grounds. A man fishing for *'opelu* leaves the fishing grounds when a *mahimahi* is seen hovering; the fishing is spoiled.

In the Fornander collection of Hawaiian writings (25, III: 184), there are some notes on the dolphin:

The dolphin is a very game fish when caught with a hook, it is a great struggler and snorts when leaping up. A large fish is a fathom and over in length, and a small fish is . . . four and a half feet. A large fish is called a *lapalapa* (large), also *ao*, having a breadth of a yard from the forehead to the mouth. Here are the names of the dolphin: *lapalapa, ao, papa'ohe* (board of *'ohe* wood).

The *Coryphaena equiselis* Linné (26, p. 137 and 40, p. 205) has, according to Fowler's figure no. 31, a rounded snout, when young. This may account for the Hawaiian *mahimahi lapa* and *mahimahi oma.* Or the Hawaiians may be correct, and one is male, the other female. Poey remarks on differences in male and female of *C. equiselis* (quoted by Jordan and Evermann, 40, p. 205).

Ma'iholu, included in the Hawaiian dictionary where the meaning is not clear: " This fish includes the e'a and the a'awa." (2, p. 398.)

Mai'i (ma'i'i'i) (tiny). Reports differ, some say it is the young of the *pualu,* others that it is a distinct fish, the *Acanthurus nigrofuscus.* Watson says the *mai'i* is about 2 to 6 inches long, and is striped vertically from dorsal to ventral, with black and white, throughout the body;

Mai'i

fins are dark. He adds that it is better eating than the *pualu,* and can be eaten both raw and cooked. It is usually broiled.

Maiko, a surgeon fish: *Acanthurus leucopareius* (Jenkins), *A. lineolatus* (Valenciennes).

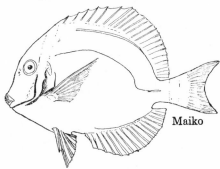

Maiko

Description: Size, from 3 to 9 inches in length, body compressed; colour, the *A. leucopareius* is brown, with two whitish bands, one from the base of the first and second dorsal spines down back of the eye to the jaw; the second band is across the caudal fin at its base. The *A. lineolatus* is coppery brown with very narrow, pale blue lines more or less from dorsal to ventral sides, a jet black spot on caudal peduncle at base of the last dorsal ray. There is a caudal spine.

Eaten broiled or raw, after the skin is removed; odour is strong; taste is preferable to the *palani*, odour not quite so strong.

A common fish.

Maka-ā (bright eyes) (*ulae mahimahi*) ; *Malacanthus hoedtii* Bleeker.

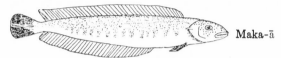

Maka-ā

Description: Length, cast in Bishop Museum is about 12 inches long; colour greyish above, yellowish white below, the grey merging unevenly; fins yellowish, except dorsal, which is salmon pink; tail fin yellow with two black streaks, bluish tinges at base; eye ringed with blue. Jordan and Evermann (40: 195) describes a specimen as "light olive green, belly silvery, side with about twenty faint short bands of the black colour, dorsal . . . a little rosy. Another source (52) says, " It is a tiny fish . . . greenish . . . delicious."

According to Jordan and Evermann's figures 75 and 76 (40: 195), the *maka-ā* and *amuka* are similar in appearance. If they resemble each other so closely, it may be incorrect to use *amuka* as an alternate name for *kahala*. But this may have been the usage in some localities. Jordan and Evermann were confused about it, evidently, for they give a name *pua kahala* (young *kahala*) for *Seriola purpurescens* (40: 183), and also for their *Carangus affinis* (40: 195).

Makapuʻu (bulging eyes), listed (59 h, 8/26/99).

Makiawa (*mikiawa, omaka* (sharp eyed)), round herring: *Etrumeus micropus* (Schlegel). In Hilo this fish is called the *omaka*, but it is not the same fish as the *omaka* described under that name in this list; the name evidently was used for two fish.

Makiawa

Description: Length, 3-10 inches; colour, steel-olive above, sides slightly yellowish, silvery beneath, fins pale, tip of snout dusky; scales have diamond-shaped, dark olive blotches.

Eaten raw, highly relished, but preferred salted for a time; also dried or baked in the *imu*, several wrapped together in a leaf bundle.

A common fish of the estuaries, plentiful in Pearl Harbour, sometimes referred to as " Ewa's well known fish." (Lahilahi Webb.) They sometimes come over the reef in large schools and are caught with nets (Kinney). If not found in a school they are not abundant, a few only being in the average haul of a bait net (Watson).

Makimaki, see '*o'opuhue*.

Makolokolopā (creeping along like a wall), or *makolokolopao*, the young of the *kūmū* or *weke*.

Makua (parent), see *kunehi*.

Makukana, cow fish or trunk fish: *Lactoria fornasini*. Hawaiian name is found only in Jordan and Evermann (40, p. 445). Not eaten by Hawaiians; in some parts of the tropics this is poisonous.

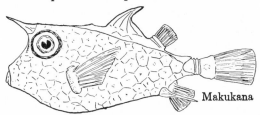

Makukana

Malamalama (light, as the light of day), see under *hilu*.

Male, or *omamale*, see *uhu*.

Mālolo, flying fish: *Cypselurus simus* (Valenciennes), *Parexocoetus brachypterus* (Richardson), *Exocoetus volitans* Linné, and others of the *Exocoetidae*. Fowler lists eight for Hawaii, and a few others for adjacent waters to the northwest. Hawaiians named four kinds —possibly other names are lost—the common *mālolo*, probably *C. simus*, about 1 foot long; *m. eheu-la* (sail-like wings), described as having reddish pectoral fins and tail, "seems of better flavour than the others," possibly *P. brachypharus*, about 9 inches long; *m. hapu'u* (tree fern), darker, its mouth broader, possibly *E. volitans*, a small species, and *lelepo* (distant sea), about 18 inches long, bluish dorsal region, long pectorals, a night flier, all others flying only in the daytime, according to one informant. The young of all flying fish are called *puhiki'i* (or *puiki*). One writer, Beckley (4), makes exception to this, considering the *puhiki'i* a separate species, smaller than the others.

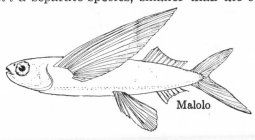

Malolo

Description: Sizes indicated above; colour, in general they are bluish dorsally, with whitish ventral surface; fins and tail sometimes whitish, sometimes touched with red.

Mālolo was much sought for food. It was cooked in *ti* leaves, but most frequently eaten raw. The *puhiki'i* were a favourite for eating raw, and combining with the fragrant *lipoa* seaweed.

Parties of fishermen went out to catch *mālolo*, surrounding a school with nets when they could. The best season for catching them is March to June. All varieties swim near the surface of the deep sea, rising out of the water and skimming the tops of the waves. Their pectoral fins serve as wings, except that they cannot move them as do birds, and therefore they do not actually fly.

Mamǔmo (*mamano, ma'oma'o* (green), *mamo, mamo pohole*) : *Abudefduf abdominalis* (Quoy and Gaimard). The name common in Oahu is *mamo*.

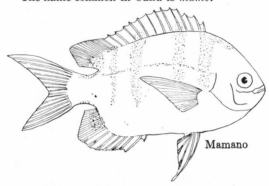

Mamano

Description: Size 3 to 4 inches, sometimes larger; body compressed; colour, white, others say pale, brassy green, with five bluish black vertical bars along sides from and through lower part of dorsal fin three-quarters of the height of the body. Kepelino gives the colours of the bars as dark, almost purple. Scales large, rounded.

The chiefs were fond of this fish; for softness, *ma'oma'o* was the best, good to eat raw or broiled.

Manaloa, see *nenue*.

Mananalo (tasty but not sweet), or *ananalo*, see *hinālea*.

Mane'one'o, see *kihikihi mane'one'o*, and *humuhumu mane-'one'o*.

Manini, a surgeon fish: *Acanthurus triostegus* (Linné)   A descriptive name of this or some other *manini* is *m. 'elekuhō* ('*ele*, dark; *kuhō*, splash made by an object or animal falling into the water). Stages of growth are: *'ōhua-liko* (tender leaf bud), transparent, size of postage stamp, *'ōhua-kani'o*, stripes appear when the fish is about a day old, *'ōhua-pala-pohaku* (stone slime), or *'ōhua-ha'eka'eka* (grimy looking) ; at this stage they begin to nibble at fine *pala-pohaku* seaweed and the skin begins to darken, *kakala-manini,* half-grown, and *manini,* adult stage.

Manini

Description: Length, 3 to 5 inches; body compressed; colour yellowish grey with reddish grey stripes; a caudal spine is present, called *kakala,* or *kala* (the spur of a cock).

One of the most popular and sought for fishes, delicious, liked by chiefs and commoners, even though it is tough-skinned and bony. The flesh is white. There are many references to its being eaten raw, and very fresh. " The table was set for them and Kalelehua went to the side of the house. In a short time she returned holding a wooden dish filled with *manini* and *lipoa* seaweed. She remarked to her attendant, ' This is a mountain place, far from the sea, yet the fish are still moving!'" (75.18). This is a fish that was never cleaned of its soft parts, except for removal of the gall bladder from the full-grown fish if it was to be cooked. One Hawaiian said, " You have never eaten *manini* until you have eaten it whole, that is, not cleaned of entrails. If cooked, it should be broiled." Sometimes the skin was removed to be saved for *palu.* When eaten raw, *manini* were usually salted first. Salting flattened down the dorsal fin, which becomes stiff and upright (*i moe kakala*) soon after the fish is taken from the water, making the fish hazardous to swallow. *Manini* was a popular fish for drying.

Of the *manini*, Mrs. Pukui relates:—

It was one of the commonest fishes of my birthplace, Ka'u, in Hawaii. Every summer the natives used to catch the spawn by the million. Large, flat lava rocks were swept clean with brooms of coconut midribs. Then the *'ōhua*, about postage stamp size, were mixed with salt and scattered to dry in the hot sun. Later they were stored for the future, or taken inland for exchange with those who lived far from the shore.

Manō, shark. Several Hawaiian names have been recorded for sharks, but it has not been possible as yet to attach scientific names to all of them, or know whether some sharks have more than one Hawaiian name. *Manō* was the name for all, but the *niūhi*, the man-eating shark, usually went by that name alone. Some Hawaiian writers have made a distinction between *manō i'a* (edible sharks), and *manō kanaka* ('aumakua sharks).

Hawaiian sharks as listed by Fowler (26, pp. 18-24) are: *Carcharodon carcharias* Linné, *Isurus glaucus* (Müller and Henle), *Alopias vulpinus* (Bonnaterre), *Galeocerdo arcticus* (Faber), *Glyphis glaucus* (Linné), *Eulamia sorrah* (Müller and Henle), *E. melanopterus* (Quoy and Gaimard), *E. commersonii* (Blainville), *E. munsing* (Bleeker), *Sphryna zygaena* (Linné), *S. tudes* (Valenciennes), *Squalus suckleyi* (Girard), *Etmopterus villosus* Gilbert, *Centroscullium nigrum* Garman, *Insistius brasiliensis* (Quoy and Gaimard) and *Enprotomicrus bispinatus* (Quoy and Gaimard).*

Hawaiian names collected are: *manō* ?, a large white shark, the largest known to Hawaiians, rare, not ravenous, *m. kihikihi* (angular), hammerhead: *Sphryna zygaena*, common in many areas, as Kane'ohe Bay, *m. hi'uka* (smite tail), thresher shark: *Alopias vulpinus* (Bonnaterre), rare, sometimes seen in Ka'u, *manō lālākea* (white fin), perhaps the *Squalus suckleyi* (Girard), never known to attack human beings (4: 82), *m. leleiwi, m. luia,* or *luhia, m. laukahi'u* (long-tailed), possibly another name for the thresher, *m. 'ula* (red), *m. pa'ele* (black smudged), *m. moelawa, m. leleiona* (see

* At the present time (1970), intensive studies are being conducted on Hawaiian sharks. The principal sharks, as to numbers, reckoned by the 1967–1969 catch, are: tiger (*Galeocerdo cuvieri*), sandbar (*Carcharhinus milberti*), Galapagos (*C. galapagensis*), reef gray (*C. amblyrhynchos*). See Albert L. Tester, "Cooperative Shark Research and Control Program, Final Report, 1967–1969," University of Hawaii, Department of Zoology.

leleliona, the sucker fish), *m. lelewaʻa* (canoe leaping), *m. pahaha* (thick-necked), *niūhi*, the man-eater or great grey shark: *Carcharodon carcharias* (Linné), sometimes called *niūhi ʻai-lawa* (man-eating *niūhi*). Of these only three are here described, the first two because they were often eaten—few others were used for food; the third was feared, respected, sought for purposes other than food:—

Kihikihi, hammerhead. " A large voracious shark reaching a length of 15 feet or more, found in all warm seas; colour grey." (40: 41).

Manō kihikihi

Lālākea, reef shark, called the Hawaiian dog-fish by Bryan (11, p. 346). Length, 4 to 6 feet long; colour, grey, with lighter coloured fins. Hawaiians call this harmless. It sometimes follows fishermen about the reef waters, perhaps looking for discarded fish.

Manō lālākea

Niūhi, " one of the largest sharks, reaching a length of 30 feet; found in all temperate and tropical seas . . ." (40: 44). A cast in Bishop Museum shows it to be dark grey, with lighter underbody, the body thick and heavy, the tail lobes almost equal. Informants (Anderson and Watson) say the eyes are luminous at night. If a glow in the water is seen at night it is wisest to stop fishing and go home for the fishing is spoiled by the presence of this shark, and it is dangerous to be near as the shark may attack the canoe (65: 10-12). An old chant (23a: 221) also records Hawaiian observation of the *niūhi's* eyes:—

> . . . the great shark
> Niuhi with fiery eyes
> That flash in the deep sea.
> Oh! Alas!
> When the *wiliwili* tree blooms
> The shark bites . . .

Kinney says that they become enraged with the smell or sight of blood. Still another says (4: 19) that catching the *niūhi* was the game of kings; it was a dangerous sport and special technique was developed for catching it. Favoured persons were given the right to fish for *niūhi*; it was tabu for others to fish for it, breaking of this tabu punishable by death.

Niūhi were not eaten but the teeth were highly prized for cutting edges of knives for domestic and warfare uses.

Sharks that attacked or ate human beings were not eaten. Those eaten most commonly were the *kihikihi* and *lālākea*, evidently the least vicious. Some other sharks were eaten by some but were tabu to others. To those who could eat them it was a day of rejoicing when one was captured. The meat was dried, after removing the skin. This was done most easily by holding the fish, or a portion of it, over hot stones of the *imu* then plunging it into hot water for just the right length of time. The flesh was cut into strips, salted and dried. After it was well dried it was broiled or cooked in the *imu* as needed. Another method of preparing the flesh was to cut it from the backbone, and cut it into strips without breaking the skin, salt it and put it into a trough-like container that was tilted slightly to let some of the brine run out. It was left overnight, rinsed to remove some of the salt, then dried, and so was ready for broiling or cooking in the *imu*. No one reports that shark was ever eaten raw, and one informant had known many shark eaters.

Hawaiians who avoided eating shark meat did so because the shark was their *'aumakua*. Such sharks were called *manō kanaka*, supposedly born of a human mother and a shark father. They were persons who took a shark form after death, and revealed their relationship through a dream experienced by the living relative. From then on relatives of the deceased were required to avoid eating shark and had to feed a special shark thought to be such a relative by taking bananas, *'awa*, and other foods to the seashore, uttering prayers to accompany the offerings, and tossing the food into the sea. In many tales the shark came near shore to take the offering, in full sight of the devotee. Shark *'aumakua* protected their devotees. If they fell into the water, at sea, or if a canoe were upset, they would be brought safely to land by the shark.

Shark stories are frequent in Hawaiian literature. One is selected from the legend of Punia (25, II: 294-308, abstracted) :—

> The hero has eleven sharks to contend against, for they are blocking a lobster hole from which Punia and his widowed mother hope to get lobsters to go with their poi. The chief of the sharks has a name, Kaialeale, and with him Punia converses. Punia tricks the sharks to get out of his way, gets a couple of lobsters and then says, " This will keep my mother and myself alive. It

was the first shark, the second shark, the third shark . . . the eleventh shark that told me what to do, the one with the thin tail. He was the one that told me what to do." When Kaialeale heard this . . . he ordered all the sharks to come together and get in a row. He looked for the one with the thin tail. " So it was you that told Punia what to do. You shall die." When this act is carried out, Punia cries out, " So you have killed one of your own kind." This trickery and taunting go on until all the sharks are killed except Kaialeale. Punia boldly enters this shark's mouth, props it open with sticks as he slips through, lives in the shark's stomach for several days, scraping his sustenance from the walls of his prison, to the great discomfiture of his host, finally tricks the shark to beach himself, and frees himself when the shore people have cut open poor, stupid Kaialeale.

Emerson (22, p. 8-10) gives a long description of the *'aumakua* and says that the shark was

perhaps the most universally worshipped of all the *aumakua* . . . each locality along the coast . . . had its special patron shark . . . well known to all frequenters of that coast. Each of these sharks had its own *kahu* (keeper) who was responsible for its care and worship. The office of *kahu* was hereditary . . . the relationship between a shark god and its *kahu* was oftentimes of the most intimate and confidential nature. The shark enjoyed the caresses of its *kahu* as it came from time to time to receive a pig, a fowl, a piece of *'awa*, a *malo*, or some other token of its *kahu's* devotion. And in turn it was always ready to aid and assist the *kahu*."

An instance of this aid was in the rescue of one *kahu* at sea. The shark *'aumakua* came and took her " upon his back to the neighbourhood of Kaho'olawe " (an island off Maui).

Teeth of all sharks were prized for their sharpness and strength. (12, pp. 27-41). The rough skin was sought as material for drum heads.

Ma'oma'o, see *mamamo*.

Maui, listed (59 i).

Ma'ula'ula (reddish), see *'ula'ula*.

Maumau, listed (13: 465).

Me'eme'e, see *iheihe*.

Meumeu (taper), name on a Hawaiian list.

Mikiawa, see *makiawa*.

Mimi (urine), see *humuhumu mimi*.

Moa (wooden block) ; *moamoa*, see *pahu*.

Moano, one of the goatfishes: *Parupeneus multifasciatus* J. and E., and probably others. *Moano* was one of the fishes the young of which were called *'ahua* or *'ohua*. Six names of varieties have been collected: *m. auki* (ti stem), *m. kea* (white) (*P. filamentosus*), *m. papa'a* (burnt), *m. ukali* (following), *m. ukali ulua* (*ulua* following), *m. ahulu* (over cooked). One Hawaiian says there are four kinds, and gives their colour notes, but knows no names for them: red with black spots; another light red, no spots; another black with yellow blotch on tail; and a fourth purplish, with yellow fin, and black tail spot.

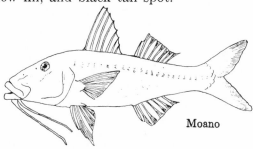

Moano

Description: Length, 6 to 8 inches, sometimes as long as 12 inches; colour, most common varieties are the red ones, tips of the scales are yellow. Scales large.

Eaten raw, or broiled in *ti* leaves. Kepelino (52) reported that *moano* was "one of most delicious fishes when cooked in *ti* leaves." A few lines of a chant mention this fish: " *Ono, ono wale mai la no ka hoi ka i'a o ke kai, A he moano kai lena, Ono! Ono!"* (Delicious, delicious is the fish of the sea, The moano of the yellowish sea, delicious, delicious!) (75.29).

Hawaiians believed that this fish ate *lehua* blossoms, which are a deep red, and derived their colour from the flowers. Sometimes this fish was referred to as *moano-nui-ka-lehua* (great *moano* of the *lehua*).

It is found in reef waters and the deep waters just beyond the reef.

Moi, thread-fish: *Polydactylus sexfilis* (Valenciennes). Stages of growth are: *moi-li'i* (little), 2 to 3 inches long, *palā-moi* (growing into *moi*) (Kauai term), or *mana-moi* (Hawaii term), 5 inches long, and *moi*, adult, average length is about 18 inches, an occasional fish attains as much as 3 feet, however. In the island of Hawaii, four stages of growth were named, *moi-li'i*, finger length, *mana-moi*, length of hand up to wrist, *palā-moi*, about 12 inches, *moi*, 18 inches or more. One informant speaks of the *moi-kumulau* (mother of litters), perhaps a variety.

Description: Length is given above; colour, cast in Bishop Museum is tawny yellow, with about twenty narrow, light brown stripes lengthwise. The colour depends on the sea bottom over which the fish is loitering, dark grey on the reefs, a lighter shade over sandy

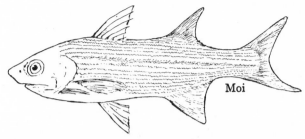

bottom.   Little schools of *moi-li'i* are seen along the shore—a soft, inconspicuous grey colour—about mid-August through October.

Moi is a fish for chiefs, and is much sought today as a delicious fish.   Formerly commoners were not allowed to eat it.   Some say it was always eaten raw, but other informants say it was salted, dried, or cooked in *ti* leaves or in the *imu*.

Oahu was noted for *moi*, and Kaena Point, and Mokapu were good spots to find it, but it was not and is not common.   It is a seaweed eater, and has favourite feeding spots.   It is hard to catch, being a great fighter.   It loves foamy, rough places in shallow water.   They often travel in large schools, and were an omen of disaster to the chiefs when they appeared in large numbers.

Mū (*mamamū* is an alternate name, according to Jordan and Evermann (40, p. 243), but no Hawaiian so far has corroborated this statement), porgy?, Australian snapper?: *Monotaxis grandoculis* (Forskal), and probably *Pagrosomus auratus* (Houttuyn).   According to two Hawaiian sources there are three varieties, but no comparative descriptions have been received.   Fowler says (26, pp. 218-219) of the *P. auratus*, " Jordan and Evermann have identified this specimen with *Monotaxis grandoculis* though only as a poor guess.   There is no question about its identity with the Australian snapper."

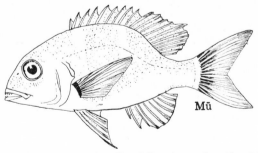

Description of *M. grandoculis*: " Reaches a length of 22 inches " (40, p. 244), one Hawaiian informant says of *mū* " about 3 feet; weight 25-30 pounds "; colour, one informant says light reddish, J. and E. (40, p. 244) says " in alcohol, greyish brown above, lighter below, margin of scales darker . . ." Cast No. 472 in Bishop Museum of *M. grandoculis* is coloured greenish white, slightly darker above lateral line, head darker, pectoral fins red, ventrals touched with red, dorsal and anal light coloured, with black areas, and red touches, caudal reddish brown. Scales large. It is about 10 inches long, evidently a young specimen.

Excellent food fish, not eaten raw, usually broiled, sometimes cooked in *imu*, but the large *mū* too large to cook whole.

The strong resemblance in form of teeth and jaws to those of man caused Hawaiians to transfer the name *mū* to the man who was sent out to get persons to be buried alive beside the body of a dead chief. (Kinney.)

Habitat is very deep water, though " the darker, smaller variety is sometimes found near the reef. They bite only at night." (Nakuina.)

Mukau; *Brama raia* (Bloch).

This seems to be a rare fish and Hawaiians consulted are unfamiliar with it. A description of *Eumegistus illustris* Jordan and Jordan (43, p. 35) is said by Fowler (26, p. 138) to " agree closely

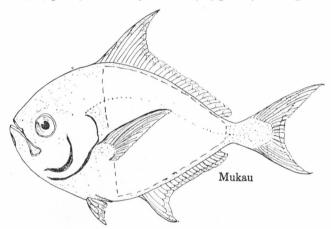

Mukau

with *Brama raia.*" The fish described by Jordan and Jordan was a solitary specimen, part of which had to be left with the fish dealer. Extracts from this description are: " Scales on head small . . . on body thick, smooth . . . those on sides much larger than those along bases of dorsal and anal fins; each ray of dorsal and anal with a series of scales . . . lustrous brownish black . . . posterior edge of caudal abruptly white, outer edges of pectorals and ventrals also white." The size is 2 feet long, weight about 9 pounds. " It was regarded as one of the best food fishes, selling at fifty cents a pound, but no one (no scientist?) seems to have seen it before."

Mukomimi, see *omilu.*

Naiʻi (*nuao*), the porpoise.

Some Hawaiians say *nuao* is the blackfish. Green (28, p. 15) calls *naiʻa* the killer whale. Some Hawaiians say porpoise was not eaten at all, others say eaten to some extent. The meat is very dark and " smells worse than shark," the odour very persistent. It was tabu to women and to those to whom shark was tabu, because it was thought there was relationship between shark and porpoise. To fishermen it was a pest for it got in their way when they were after better fish; sometimes used as shark bait.

Nakea (whitish), see *'o'opu.*

Namu (unintelligible), listed (59 i).

Nana, listed (59 f).

Naniha, perhaps also *nanihu*, see *'o'opu naniha.*

Napili (clinging), see *'o'opu.*

Nawao (of the mountains), see *'o'opu.*

Na'u (yellow), listed (59 h).

Nehu, anchovy: *Stolephorus purpureus* Fowler. Three Hawaiian names have been recorded, *nehu kulani* (of the heavens, royal), *n. pala* (yellowish), *n. maoli* (indigenous). Kamakau (47, Chap. 9), speaks of the " yellow *nehu* and the common variety."

Nehu

Description: Length, ¼ to 2½ inches; colour, bluish with silvery band on sides from head to tail; scales large.

Eaten raw or dried.

Kamakau in the reference above says that the *nehu* " filled the lochs from the channel of Pu'uloa (Pearl Harbour) inland to the Ewas; hence the saying . . . " The sea that blows up the *nehu* fish, the sea of Ewa that blows them up in rows until they rest in the calm— great Ewa of La'akona." Great schools appeared at other parts of the islands too, notably at Waihe'e in Maui (48) and at Hilo, where it was said, " Hilinehu (January) is the month for *nehu*. fish. This fish relieved the fish hunger of Hilo." (75.35.)   Both Japanese and Hawaiians use this as a bait fish for *aku*, tossing it into the water in quantity.

Nenu, see *nunu.*

Nenue (*nanue, enenue, nenuwe, manaloa*), rudder or pilot fish: *Kyphosus fuscus* (Lacépède). Hawaiian varieties: *nenue pa'i'i'i, n. paki'iki'i* (flat), (these two may be identical, they are said to be the tenderest), *n. kea* (white), (blue with white underbody), *n. uli* (dark) (yellowish, called the " queen " *nenue*), *n. elele* (messenger), difficult to catch because it leaps over the nets; tougher than others. This and the *uli* are reported from the Ka'u district, also Kane'ohe; probably every-

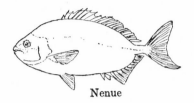

Nenue

where. All through the South Seas of Polynesia *nanue*
is the name used (Anderson). In Kane'ohe region,
nenue stands for the young, *enenue* for the adult. Not a
popular fish in Kane'ohe, where fish were and are
abundant, and the choice wide.

Description: Length, about a foot to 18 inches or even 3 feet;
colour, Fowler (26, p. 222) says "remarkable for their very great
variation in colour. All the large examples are either partly or
entirely brilliant lemon-yellow, one with piebald appearance inter-
mediate with the dark or dusky form. Most small specimens are
dusky." However, in Bishop Museum are two casts. One, No. 369,
is of a large specimen, about 2 feet long, which is greenish grey all
over, except the snout, which is yellowish. Scales are edged with
brown. The other cast, No. 37, is piebald, front and middle parts of
body are covered with two irregular blotches of clear yellow, the
rest of body is covered with two irregular blotches of clear yellow,
the rest of the body grey green; fins all dark, except the pectoral,
which is piebald. It measures about 21½ inches. The variation in
colouration must be even greater than Fowler had opportunity to
observe. Watson (inf.) is skeptical of Fowler's "yellow" colour.
He says there may be two or three yellowish individuals in a thousand,
such individuals called the *makua* (relative or protector; corresponds
to "queen") of the *nenue*. Kondo says that practically every school
has one yellow individual, but there are seldom many. Watson says a
fisherman is lucky to get five among a hundred. Nakuina (inf.) says
nenue are dark greyish; Kepelino (52) says, "Dark, greenish grey,
with dull gold stripe running from gills to tail; its belly whitish."
The dull gold stripe is mentioned by him alone. Scales are small,
difficult to remove.

This fish is one of those called delicious. Some say it is the best
raw fish. If cooked, it is best when wrapped in *ti* leaves, then broiled.
It is good for *palu*. Pa'ahana Wiggin says, "A choice fish in Ka'u.
It feeds on *limu nanue* and *limu kala*. (Both of these seaweeds
evidently take their names from the fish that eat them.) We call it
i'a papa (*papa*: flat reef area near shore) because it remained inside
the reef close to shore. When very fat only the gall bladder
was removed. The entrails were saved to make a relish. The
head was chopped into small pieces and added to the entrails and salt,
kukui nut, and chili pepper added, as usual." Kinney says it was one
of the most popular fishes with the Hawaiians, reserved for the chiefs
in the old days. The odour is strong.

Newenewe (feeble), listed (59 f), perhaps another spelling
or a misspelling of *nenue*.

Nihipali (sneak along the cliff), listed (13: 478).

Nihokomaka (sharp-toothed), listed (59 h).

Nohu (sand burr) (*hahili*). Several varieties of the *Scor-paenidae* are called *nohu*. The *Merinthe macrocephala*
(Sauvage) is given the name *'o'opu kai nohu* by Jordan
and Evermann (40: 461). I do not know how widely
this name was used; most of the *'o'opu* are gobies. I
think that borrowing the term for this *nohu* was to
distinguish it, as we would say a wolf-like dog. These
Hawaiian varieties have been noted: *nohu pinao*
'(dragon-fly), *n. omakaha* (streaked), and *n. po'ola'au*
(wooden head). One casual description by an informant
is that they are much the same in looks, some have a
little yellow on the fins, and another that the general
colouring is deep bluish, spotted with yellow, and
along the edges of the fins are rows of red dots.

Besides the *Merinthe*, the following have been
identified as *nohu: Scorpaenopsis cacopsis* (Jenkins),
S. gibbosa J. and E., both of which are *n. omakaha*.

Nohu

Description: Length, about 1 foot and a half, some smaller, as 6 to 9 inches for *S. gibbosa* (40: 469); colour, it seems intent on camouflage, all colours being broken up into mottlings and surfaces broken and bumpy. Little protuberances from the skin resemble bits of seaweed. J. and E. descriptions (40, pp. 461-470) vary for the different species: rusty, reddish brown with touches of white or pale rose (*S. cacopsis*), or flesh colour mottled, breast and belly yellowish (*S. gibbosa*), or dark purplish brown, excessively mottled (*Merinthe macrocephala*), or very pale brown (in alcohol), whitish beneath, with broad deep brown bands, very much banded (*Pterois spex*). All have large heads, and spines on the head and the gill shield, from the eye region to the dorsal spines.

The skin is coarse, the scales large.

In spite of its forbidding appearance this fish is good to eat and sought for as food. It is always cooked. The Chinese in Hawaii are especially fond of it.

Hawaiians noted its habit of making a snoring noise, and also that it follows the man-eater shark, the larger varieties being deep-water fish. It was believed by some that the *nohu po'ola'au* laid the eggs that hatched into sharks.

Noinoi (to beg), listed (59 i).

The only other mention of this fish is by Henshaw (36, p. 125) :—
While following its prey on the broad ocean the *noio* (tern) is of much service to the Hawaiian fishermen, and acts as his pilot; for its presence in numbers in a given spot marks the where-abouts of shoals of *noi*, a long silvery minnow, and there also is sure to be found the *aku* ...

Nopili (cling), see *'o'opu*.

Nuao, see *nai'a*.

Nuhu, see *nunu*.

Nukuheu (snout fine; hair-like growth), listed (59 i).

Nukumomi (pearly snout), see *ulua*.

Nukunuku (beaked), butterfly fish: *Forcipiger longirostris* (Broussonet).

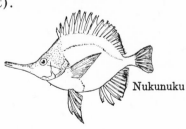

Nukunuku

Description: Size, about 6 inches, colour, brilliant yellow, upper part of head to dorsal fin black above, white below, fins yellow except caudal which is colourless, large black spot on anal fin, the last spines of dorsal and anal fins touched with black.

Good to eat, though there is little flesh; broiled.

Nūnū (*nenū*), trumpet fish: *Aulostomus chinensis* (Linné).

Nūnū

Description: Length, usually about 12 inches (52); J. and E. found a specimen 19.5 inches; an informant, Watson, remembers seeing a fish about 4 feet long. Colour, subject to great variation (40, p. 115), lemon yellow, or light brown, with rosy-brown, longitudinal stripes, head pale rosy, fins all pale rosy. Kepelino (52) says entire colouring a dark, yellowish blue, fins yellowish; Mrs. Pukui says blue.

Eaten broiled, or wrapped in *ti* leaves and broiled, or dried.

Lives in shallow water, often found with the *'aha.*

Oala (bucking motion), no data except from Andrews-Parker dictionary (2): a class of fishes that feed on *limu* or moss, like the *kala, nenue,* etc., so named because they toss themselves about when feeding.

Oama, see *weke.*

'Ohe (bamboo), listed (59 h).

Ohio (prying motion), Andrews-Parker dictionary (2) says, same as *ohi'u* (prying motion), a fish, and under *ohi'u,* no definition, however under *ohi'uhi'u* (repeatedly prying), there is the definition, "the blue *uhu* (*Julis lepomis*); so called at Kawaihae" (Island of Hawaii).

Ohi'uhi'u (repeatedly prying), see *uhu.*

Oholehole, same as *aholehole* (59 a).

'Ohua (seedling), (*'ahua,* especially the young of the *kawakawa, aku* and *moano*), the term is applied to the spawn of such fish as *uhu* (*'ohua palemo*), *manini* (*'ohua liko* or *o. kani'o,* etc.), *puwalu, kupou, hinālea, kala.* Some names for *'ohua* have survived, though the name of the fish at its adult state is lost to memory, possibly some *'ohua* were not identified with their adult state. Some of these are: *kukaepua'a* (hog excrement) mentioned in legends, *li'i* (little), *lipoa* (fragrant seaweed), *niho-nui* (big tooth), *pa'awela* (like the *awela*), *unahi-nui* (big scale), *makali'i* (tiny).

Fish spawn were caught in large quantities, with hand nets and scoops. They were cleaned by working them with salt, and eaten raw or cooked in *ti* leaves; highly prized (4, p. 14).

Mrs. Pukui says, " Sometimes these fish were caught before they had escaped from the transparent bag in which they had been hatched. Mrs. Makahonu lived in Ni'ihau and used to see the whales come at spawning season of the *'ohua* (*manini*) to feed upon the young fish. Fishermen seeing whales so often in proximity of these " bags " of fish, thought that the whales had blown the bags out from their own vast interiors. To this day such bags of spawn, large enough to fill a bucket, are called *upe* (or *hupe*) *kohalā* (mucus from the whale's nose). Ambergris was called by the same term.

The legendary source of the name *kukaepua'a* for one of the *'ohua* is a tale of the demi-god, Kamapua'a, one of whose forms was that of a pig. He was surfing at Kamaka-loa, a beach in Puna, Hawaii. Kamapua'a, always mischevious, drove a school of these small fish up to the shore. Pele and her sisters caught and ate some of the fish, whereupon Kamapua'a taunted them, " Ho! You are eating hog excrement!"

Ohune (small, a fragment), or *ohuna*, see *'o'opu*.

Oie, a name on a list.

'O'ili (make a sudden appearance): *Cantherhines sandwichensis* (Quoy and Gaimard). With the exception of the *Alutera monoceros* (Osbeck), the *loulou* (an indigenous palm), the fishes of the *Monacanthidae* family (file fishes) are called *'o'ili*. Names reported are: *'o'ili uwiwi* (squeaky sound), (also called *uwiuwi*, or *oeoe*) (protruding motion, of a goose's neck): *Pervagor spilosoma* Fraser-Brunner; *'o'ili lepa* (*'o'ilepa*) (flag bearer): *Cantherhines sandwichensis* (Quoy and Gaimard).

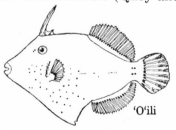

'O'ili

Description: Length, 2 or 3 inches up to 24 inches; colour, *'o'ililepa*, black with reddish-yellow fins; *uwiuwi*, yellow with black dots on body, diagonal stripe on head, tail orange, dorsal and ventral pale yellow. There are other colour varieties. Skin rough, scales very small, some kinds have two spines on the caudal peduncle. The chief difference between the *'o'ili* and the *humuhumu* being the distance between the dorsal spines, the *'o'ili* having one or two high spines set far in advance of the soft dorsal fin.

According to an informant for J. and E. (40, p. 421), this fish (*'o'ili uwiwi*) comes occasionally in great numbers but otherwise is very rare. The natives believe its appearances to prophesy the demise of some great personage, such as a king or a chief. In the spring of 1944, they were observed in great numbers, millions of them all along the beaches of southeastern Oahu. They seemed to be already dead when they floated ashore and piled up on the beaches. This continued for weeks. Many persons tried to lay the cause to wartime occurrences at sea, but it is likely that the cause had nothing to do with the war, in the light of what Jordan and Evermann recorded above.

Evidently they were sometimes eaten. In 1862, Kepelino says (52) " the skin is peeled off and thrown away, and then they are eaten raw or broiled. This is a fleshy fish, not many bones. (Those washed ashore in 1944 were exceedingly compressed in form, about 3 to 4 inches long, and the flesh must have been of trifling account.) It is a fish used in the idol worship of worthless people in times past and some are still indulging in the practice." Mrs. Pukui says, " In Ka'u they are sometimes blown ashore in times of strong gales, gathered for fuel, as they have little food value. In Oahu they were eaten. The *'o'ili uwiuwi* makes a noise such as the little pig made all the way home, hence the name."

'Ō'io, ladyfish, bonefish: *Albula vulpes* (Linné). Stages of growth are: *nehu*, name evidently borrowed from the *nehu* fish; *pua 'ō'io*, finger length; *amo'omo'o* (like small strips of unfinished tapa), forearm length; *'ō'io*.

'Ōi'o

Description: Size, adult about 2 to 3 feet long, body roundish, head flattened on top; colour, silvery. Jordan and Evermann say (40: 55), " brilliantly silvery, olivaceous above, back and sides with faint streaks along the rows of scales." The fins are neutral in colour, transparent, touched with grey along edges. In Bishop Museum cast No. 119, the head is more smooth than is indicated in the accompanying sketch taken from Jordan and Evermann's figure (40, fig. 9). Scales are large.

This is an exceedingly popular food fish, flesh is delicious, white; liked raw when its many fine bones are supple and slip down the throat without any trouble; often eaten " lomied " with *limu kohu* (a seaweed). It is seldom cooked for the bones then harden. In modern times it is used by Orientals for fish cake, a preparation of raw flesh scraped off the bones. The resulting mash is used in various ways, sometimes merely salted a little, rolled in egg and flour and fried, sometimes dropped by small spoonful into a thin bouillon or clear fish soup or broth. Hawaiians today sometimes scrape the flesh from the bones after letting the flesh stand 3 to 4 hours when it loses some of its firmness.

" '*Ō'io* travel in schools, and are caught with nets or lines at their feeding grounds, sandy places where seaweed grows, though not in brackish water. The nose is hard and bony, and with it the fish churns up the sandy spots in search of tiny crabs and other sea animals. It is also fond of seaweed. The '*ō'io* from Keahi, between Kupaka and Pu'uloa (on Oahu), was noted for its fragrance which was like that of the *lipoa* seaweed, and it sold much faster than '*ō'io* from other localities." (Pa'ahana Wiggin.)

Okuhekuhe (big mouthed), '*o'opu*.

Olale or *olali*, see *hou*.

Olani, see *hou*.

Olelepa (flag bearer), listed (59 a), perhaps a misspelling or variant for '*o'ililepa*.

Olola, according to the Andrews-Parker dictionary (2, p. 487), "a species of small mullet; a fish resembling the *puhiki'i*, called the parent of the *mālolo*."

Oma, a variety of *mahimahi*.

Omaka. This name is used for two fish. One is the Hawaiian herring, which is here listed as the *makiawa* (Oahu name; in Hilo, *omaka* is the name used). The other fish called *omaka* is, according to J. and E. (40, p. 283), *Stethojulis axillaris* (Quoy and Gaimard).

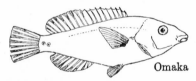

Omaka

Description: Size, about 6 inches, body moderately compressed; colour, olive, with irregular white areas, fins golden with pink dots, two or three black spots on caudal peduncle, golden spot on base of pectoral fin. (40, p. 283.)

Omakaha (streaked), a variety of *nohu*.

Omale, see *uhu*.

Omilu (*omilumilu*) (insignificant). Another name is *muko-mimi*, perhaps for one species only.

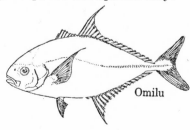

Omilu

According to Nakuina, the *omilu* is "a type of *ulua*, mouth resembles that of mullet, but the striking difference between *ulua* and *omilu* is that *ulua* has teeth, *omilu* has not." This would indicate that the *omilu* are *Carangoides* of Jordan and Evermann (40, pp. 198-200), of which three species are given, Fowler's names for which are almost identical with Jordan and Evermann, namely, *Carangoides ferdau* (Forskal), *C. gymnosthethoides* Bleeker, and *C. ajax* (Snyder). Jordan and Evermann designate *C. ferdau* as *omilu*, and it seems likely that all three may take that Hawaiian name. Jordan and Evermann say of the *Carangoides* (40: 198), "teeth persistent, all small in villiform (hairlike, lying like the pile of velvet) bands on jaws . . . lateral line scarcely arched in front, body oblong, not much elevated; otherwise essentially as in *Caranx*." Nakuina says of them, "size, about 2 feet long, colour similar to *ulua*, bluish or black, with yellow face."

Notes from Jordan and Evermann's descriptions of these fish are as follows: " *C. ferdau*, silvery, dark blue reflections over upper portions of back and head, several small lemon-yellow coloured spots with dusky centres on side . . . soft dorsal and anal blue, lower portion . . . golden with blue outer margin, ventrals white with bluish shade, pectoral transparent, with golden shade; caudal bluish and golden . . . (40: 199). The spots indicate the name *omilu kukaenalo* (freckled). *C. gymnostethoides*, nearly plain olive . . . recorded from Johnston Island . . . (also in Hawaii?) (40: 199). *C. ajax*, silvery, darker above, indistinctly marbled with dusky along back . . . dorsal fin with a dusky margin (40: 200). This differs from *C. ferdau* markedly in having no spinous dorsal."

Jordan and Evermann also attach the name *omilu* (*omilumilu*) to *Caranx melampygus* Cuvier (40: 192-193), which they describe as "teeth small, sparse, in a single row in each jaw . . . general colour silvery, dusky above, lighter below, upper parts with numerous small black spots, intermingled with numerous bright dark blue spots, interspaces with . . . golden reflections; golden band along the scutes; soft dorsal, anal and a narrow area along their bases bright ultramarine . . . pectoral and caudal dusky . . . ventrals dusky with dark blue . . . It is more common in Samoa . . . and as a food fish it is superior even to the *ulua*."

Eaten raw or cooked in any way, preferred by some to its relative, the *ulua*.

Oniho (having teeth), known only from the Andrews-Parker dictionary (2, p. 494) : "a deep-sea fish of the *uhu* family."

Ono (good to eat), a large mackerel type of fish: *Acanthocybium solandri* (Cuvier). The term *ono malani* (pale) has been recorded.

Description: Size, 5 or 6 feet long, body roundish; colour, steel blue, back darker, colours differ somewhat in individuals, for brownish or black stripes are sometimes evident (age difference?) and colours

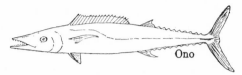

Ono

change according to place where the fish is—camouflage. Whitish underneath. Teeth and jaws very strong—" can bite anything." (Nakuina.)

A choice food fish, liked cooked only, broiled, or salted, dried, baked in the *imu*, or wrapped in *ti* leaves and placed on top of the taro mound of the *imu*. Kinney says it was eaten raw also.

This fish is not abundant; it is usually found in channels between the islands, sometimes travelling with the *mahimahi*. J. and E. say (40, p. 176), " The *ono* was said by the ancient Hawaiians to be the parent of the *'opelu* (mackerel)," and it may be added that parent is often given as the meaning for the word *makua*, which has a wider meaning than direct parent, that is, relative by blood or marriage, or even a person who accompanies in a protective role.

'O'onui, see *kupīpī*.

'O'opu. These are fishes included in the families *Eleotridae* and *Gobiidae* of Fowler (26, pp. 388-419), who lists five species of *Eleotridae* definitely Hawaiian, and twelve species of the *Gobiidae* from Hawaii.

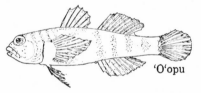

'O'opu

According to Jordan and Evermann (40, p. 478), they are " carnivorous, mostly of small size, living at the bottom near the shore . . . some inhabit fresh waters and others live indiscriminately in either fresh or salt water; many of them bury themselves in the mud of estuaries. Few of them are large enough to be of much food value." Hawaiians were very fond of them nevertheless. Mainland (62) has reviewed the species found on Oahu.

'O'opu were most plentiful in and at the mouth of the larger streams, therefore at Waipi'o valley, Hawaii, and in the several streams of Kauai. The following notes by Hawaiians tell of popularity and distribution in the old days. *Hinana* (spawn) were especially popular as dainty food.

By the mouth of the river of Waimea, Kauai, was a multitude of men and women along the banks, for those were good days in which to catch *hinana* in nets. The fish were as plentiful as rubbish in that land when the *hinana* season came. The natives there call it " *ke i'a ili kanaka o Waimea* " (the fish of Waimea that touches the skin of man) . . . (75.51).

The *hinana* was a fish of which the natives of Waimea and thereabouts were so fond that they hardly shared with others. Even though a neighbour stood by while one cleaned her fish, freshly caught, there was no raising of the head in recognition, or a word or offering to share. (Offering to share would have been obligatory if a greeting were exchanged). *Hinana* was *i'a pi ia* (fish stingily regarded). There were people so lucky in fishing that they were said to have skins like Ku'ula ('ili Ku'ula). If there were such persons in a locality only they were allowed to dive into the water with *hinana* nets. No others went into the water at that time, for that would counteract the influence or *mana* of the diver. If there were only one such person she had to go alone. Strangely, all the *'ili Ku'ula* people I knew were women.

This is how the *hinana* fishing was done near where I lived. One end of the net was tied or held fast on the bank of the stream. An *'ili Ku'ula* took the other end and swam to meet her companion from the opposite side. When the nets met they were made fast and then the two dived down to the bottom of the nets to set them right. That was all they had to do. The men then came to lift the bottom of the net to the surface. The net resembled a huge hammock, weighted with millions of the choice *hinana*. To prevent the net from tearing the fish had to be scooped out. The *'ohana* (relatives) came with calabashes, buckets, baskets, any containers, and all were filled. (Mrs. Makahonu Naumu, inf.).

The spawn, *hinana*, a very popular food, were gathered in vast quantities in certain areas. Even today the coming of this fish is worth talking about. (1940) An informant from Waimea, Kauai, says that the well-known fish of the land has appeared (May). This fish was well liked from the time of our ancestors. " *'Ai wale i ka hinana, ka i'a kaulana o ka 'aina.*" (Eat freely of the *hinana*, the well-known fish of the land.) (75.3.)

Kept in ponds, along with other fish, they were an addition to the food supply. Mounds were sometimes built in deep fish ponds, and taro planted on the mounds. Fish such as *awa*, mullet, and *'o'opu* were kept in the surrounding waters. They fed on the taro rootlets, and leaves and stems that fell into the water, also the mosses that grew there. (Kaauwana Aukai, inf.).

Ka'ahumanu owned a pond in Waipi'o Valley and devoted it to *'o'opu* entirely. " On the west side is a large pond, Muliwai. On the northwestern side of Muliwai pond are two big taro patches, Kanuniho and Umi, the largest patches in Waipi'o. Back of these is a pond for 'o'opu fish, owned by Ka'ahumanu." (Source lost.)

Beckley (4, p. 8) describes the method for catching large numbers of fish at the beginning of the rainy season. A platform was built across a stream just under the surface, where the *'o'opu* would be because the water would be clearer there.

" These *'o'opu* are highly prized as they have a very delicate flavour, from, it is supposed, having fed on the fallen flowers of the Eugenias (*lehua* trees), which always line the banks of mountain streams; they are known as *'o'opu-'ai-lehua* (*lehua*-fed *'o'opu*). The *'o'opu* breed or are hatched in salt water and the young fry ascend the streams to live and grow. The young . . . are scarcely larger than maggots . . ."

Fornander has collected from Lemuel K. N. Papa, Jr., a legend of the *'o'opu* god (25, vol. II: 510-515), Holu. In the worship of this god, two people (keepers) were charged with receiving offerings from the people to lay before the god, the suppliants asking in return that the *'o'opu* be fat and numerous. Prayers accompanied the offerings. The image of the god, the place of worship and the proper offerings are described by Papa. After the ceremony, the people return to their homes, confident that the god " would not sulk and render the *'o'opu* lean." Then came the time to construct a *hā* (watercourse or trough) for catching the *'o'opu* at times of freshet, the method differing somewhat from that described by Beckley (4).

Concerning a running contest: " Let the goal be Waipi'o Valley. The fat mullet of Paka'alana, and the *'o'opu* fish of Hi'ilawe Falls that feed on the *lehua* blossoms, these shall we bring alive and ,lay in a long wooden dish (*pa olo la'au*) before the young chief, Pu'ula, the judge of this contest. The first to get back from Waipi'o with the fish and describe the valley, shall be the winner . . ." (75.23).

Evidently they were introduced to Ni'ihau: " Here are new things in Ni'ihau, *'o'opu* fish and shrimps. There are no streams on this island, only springs . . . The *'o'opu* is like the *'o'opu nakea*, the shrimps like those of Ko'olau on Maui. The *'o'opu* are large, about seven inches long . . . One can get all he wants of shrimps and *'o'opu* so that there is no need to go to Kauai to satisfy longing . . ." (75.10).

The size of the spawn, " scarcely larger than maggots," varied somewhat according to locality and species. They were about as long as a postage stamp, and grew to about hand length (7 or 8 inches), some individuals reaching a foot or more.

Both *hinana* and full-grown *'o'opu* were usually wrapped in *ti* leaf bundles, then cooked over the coals. In Kauai ginger leaves were sometimes favoured for their delicate fragrance. *'O'opu* were also eaten raw or dried. (Beth Kahalewai, inf.).

Some *'o'opu* were used ceremonially, as will be noted for the *nopili* variety.

Hawaiian names for varieties were chosen because of head form or colour markings. They may be grouped as follows:—

1. *'O'opu kai* (sea *'o'opu*).
    ohune (fine).
    pao'o, also called *pano'o*.
    piliko'a (cling to coral) and *po'opa'a* (hard head)
        (not true *'o'opu*, discussed under their own names).

2. *'O'opu wai* (fresh water *'o'opu*).

> *hi'ukole* (raw tail), also called *hi'u-'ula* (red tail), or or *alamo'o* (lizard like).
>
> *nakea* (whitish), or *nokea*.
>
> *naniha* (avoidance).
>
> *nawao* (of the mountain), or *ki'ahamanu* (upland haunts of birds).
>
> *nopili* (cling), or *napili* (cling), or *'ai-lehua* (eat lehua).
>
> *okuhe*, or *akupa*, or *oau*, or *owau*, or *apohā* (bubble maker) or *okuhekuhe*.

3. *'O'opu* of brackish waters.

> *kani'o* (striped), or *aualaliha*—like fresh water as well.
>
> *kauleloa* (long penis).

Some fish of the *'o'opu* type that have been introduced are called *'umi'umi* (barbeled), another *i'a pākē* (a Chinese fish), and *kui* (spike). The *'o'opuhue* (big bellied) is the Tetrodon and is considered separately, following the true *'o'opu*.

'O'opu kai (sea) :—

> *ohune* (fine) : *Bathygobius fuscus* (Rüppell).

'O'opu ohune

Description: A small *'o'opu*, 4 or 5 inches long; colour, very dark, with black marblings and brown edges to scales, fins finely barred with brown; an example from Waialua, marked with pearly blue spots on a ground of light and dark olive, and reddish olive; others show a series of dusky blotches along middle of side. (40: 484). Scales large. Mainland (62, pp. 36-39) says:—

> Exceedingly abundant in Hawaii along any shore where there are rocks or broken coral to form protection, sandy shores along the reef, also estuaries . . . however, not above the tidal line. Along the shore the fish may be seen actively swimming about along the bottom exploring the undersides of rocks and coral, in short, jerky movements. When frightened they dash under the nearest rock for safety, and after a short stay usually dash on to another rock. Probably the most abundant of all the shore fishes.

Pao'o (*pano'o*). This seems to be an inclusive name for several varieties closely related, according to Hawaiian judgment, varieties which they distinguish as follows:—

> *lehei* (leaping), a swift leaper; dark body, speckled with white.
>
> *kauila* (*kauila* wood—a reddish black), reddish and striped; a *limu* (seaweed) eater, prefers to be near shore.
>
> *puhi* (eel), likes to stick its head out of the holes and look about like an eel; dark skinned.

maoli (indigenous).

moana (ocean).

'O'opu pao'o

All *pao'o* love the rough seas and rocky coasts, and love to leap from pool to pool. They are eaten dried or cooked in *ti* leaves, after being salted to taste. Fishermen often popped a live *pao'o* into the mouth when fishing but seldom ate them raw with *poi* for the flesh becomes slightly bitter after the fish dies. It was often used as bait. The *pao'o* figures in the legend of Hainakolo (75.29) :—

> The *Pao'o* fish gave such a comical answer that the bright-eyed *Kole* and the bright coloured *Hilu* . . . laughed. "This is what I do. I stay in a sea pool to quiver and move my body around. When I have lost my interest and desire for that sea pool where the seaweeds grow, I leap into another sea pool for fresh seaweeds, for the fragrance of the *lipoa* seaweed, the passing in and out of the cool sea water . . . I delight in sporting there with my slim body . . ." The *Pao'o* chanted thus:—

O Pao'o, o Pao'o keia	This is Pao'o, Pao'o
O ka kupa, ka eu kela,	That rogue, that mischief maker
Halahalakau i ka limu lipoa	That rests on the lipoa seaweed.
O ka pa keia, o ka pa kela,	A nibble here, a nibble there.
Lele au la, lekei au la,	I leap, I jump,
Lele aku au i kipuka kai nui	I leap into the large sea pools,
Lele aku au i kipuka kai li'ili'i,	I leap into the small sea pools,
O ka miki keia, o ka lawe kela,	Poking this, taking that.
Miki ke kai malo'o kahaone,	The sea ebbs, the sand is dry,
Ku ka halelo i ka 'alā,	The waterworn rocks are exposed,
Kike la! kike la!	Striking, striking.
Kike ka 'alā, uwe'ke mamane,	The rocks strike each other, the mamane tree weeps.
Ha'a e! Ha'a!	Dancing, dancing,
Ha'a ka hi'u o ka i'a pewa nui.	The tail of the fish moves as in a dance,
Halo e! Halo!	Fluttering, fluttering.
Halo na pihapiha o ka i'a.	The fins of the fish are fluttering,
Oni e! Oni!	
Oni kakaihi o Pao'o.	The Pao'o moves in joy.
E o mai 'oe, e o.	Answer, o answer me.

The *pao'o* fish was used by priests in sorcery practices, to rid a person of infatuation.

One large-scaled *pao'o* is identified as *Chlamydes cotticeps* (Steindachner), a sketch of which is included here as fig. 85.

In Bishop Museum is a cast of a "*pao'o kauleloa: Gobiichthys papuensis* (Val.)" and the note is added, "vicious, kill off all trout."

("Trout are introduced, and survive in the streams of Kauai," says Hosaka (37, p. 63).) J. and E. (40, pp. 486-487) describe this fish as "greyish-olive, faintly netted and barred with darker, and with a median dark stripe . . ." The "prevailing colour greyish olive washed in the largest specimens only with bright orange." The scales are not as large as in the *ohune*. Length between 5 and 6 inches.

'O'opu wai: *hi'ukole* (*hi'u-'ula*, or *alamo'o* (Hilo name)). This is a freshwater or mountain *'o'opu* with a pinkish or reddish tail. "It is a sign of bad luck to find one in a net when fishing for other fish for it keeps other fish away and must be thrown out of the net with an exclamation of disgust if one expects to be successful with the catch." (Mrs. Pukui.) A story of this fish was heard in Hilo, in 1930, by Mrs. Pukui:—

> A woman was fishing for *'o'opu*. She called out, "O Alamo'o, come and fill my basket until your tails stand upright." Her friends with her were surprised and disgusted to hear her calling out to a "*mo'o*" (lizard), an animal regarded with fear by Hawaiians because of its powers of evil. Her basket was filled, her friends were not as lucky. But when she emptied it at home a pinkish lizard ran out from among the fish. The woman screamed and dropped the basket. (This fish is *kapu* to many Hawaiians because of their belief that it is related to the *mo'o* gods.)

nakea (*nokea*—Hilo term): *Chonophorus guamensis* (Valenciennes). Mainland (62) names this fish *Awaous guamensis* (Val.). It has white streaks and is speckled; colour is dark olive; length sometimes as great as 12 inches. A favourite freshwater *'o'opu*, it is eaten wrapped in *ti* leaves and cooked over coals, or raw, after being salted for 12 hours. The young are caught in close-meshed scoop nets, eaten raw or cooked. *Nakea* spawn at the mouths of streams and swim upstream as far as possible. By June or July they come downstream, especially during heavy rains, "*Ua ho'opala ohi'a*" (the rain that softens the *ohi'a*). The ripened fruit of the *ohi'a* (mountain apple) drop into the streams and are borne downstream. All freshwater *'o'opu* are said to eat the blossoms of the *ohi'a lehua* that drop into the streams, hence the name *'ai-lehua* (Kinney). Mainland (62, pp. 48-50) notes that *nakea* on Oahu have a black blotch at the base of the caudal fin, dorsals and caudal light yellow or tan, each with four to seven wavy, dark brown, olivaceous or black transverse lines, other fins dusky olive to brown. He says it favours pools of fast flowing water in the upper valley, but is rarely found in the estuary (except when spawning, no doubt). It clings to the top of a boulder by means of its ventral disk. When food washes by it leaps toward it. In deep pools it may be seen crawling slowly along the bottom looking into crevices between the rocks as it moves. When frightened it dashes under the nearest rock, later emerging cautiously and clinging to the under side of the rock.

naniha: According to Kinney this *'o'opu* lives close to the stream edges where the water is fairly calm. "Poor breeders, not much eaten or relished."

nawao (Hilo name; or *kiahamanu*) : A common *'o'opu* fish of the mountain streams.

nopili (Kauai name, elsewhere called *napili*) : *Sicyopterus stimpsoni* (Gill) ; one of the *ai-lehua 'o'opu*. It can climb up a vertical stone jar or wall by moving slightly its suction disk, first on one side then the other (Mrs. Pukui). Mainland (62, pp. 32-35) gives the colour as velvety dark olive grey or nearly black; in adult males the colour more uniform and deeper; sides without transverse marking; first dorsal same colour as body, other fins edged with cream; females lighter in colour, belly blue-green to blue-grey, trunk crossed by eight to twelve olive-black transverse lines, fading ventrally, first dorsal fin faintly blue-grey, crossed with three smoky lines . . . other fins smoky, some with blue-green dots . . . Intensity of colouring varies greatly for it tends to assume the colouration of its surroundings. Size, up to 7 inches. Its habits seem similar to the *nakea*. Kinney says the *nopili* are most abundant in the higher reaches of streams, and there outnumber the *nakea*, *ohune*, and *naniha*. Kauai and Molokai are the most favoured islands for these *'o'opu*. On Kauai they are probably in all streams. The largest are found in the Wainiha, Hanalei and Makaweli streams. The *nopili* of Hanakapi'ai valley have a distinctive form, the body thicker and shorter. This characteristic has been noted in meles, *"Ka 'o'opu peke o Hanakapi'ai"* (The shortened *'o'opu* of Hanakapi'ai).

'O'opu nopili

The *nopili* was greatly relished as food, and also a favourite fish with the priests. As the *nopili* clings, so will luck. It was used in the *mawaewae* (weaning) ceremony for the first-born, so that blessings might cling to the child. It was also used in house-warming feasts, so that luck would cling to the house. Mrs. Pukui tells of a ceremony described to her several years ago :—

> A woman went to a *kahuna* to get help in "attracting success" to an undertaking. He asked her to bring him the following objects that had the *pa* (to touch) sound in their names: *pamakani*, a native hibiscus, and *naupaka*, a common beach plant, so that she would touch what she aimed at getting, and bring also a *nopili*, so the *pili* in its name (*pili*: cling) would ensure the continued success of the enterprise. *'Awa* of the *hiwa* variety was demanded also, a plant used in most ceremonies. The *kahuna* prayed for more than three hours over these objects. The woman was then given the *nopili* fish to eat and the plants to keep where they would not be defiled by careless handling. She obtained her object, so she reported, and she kept the plants until the occasion was well in the past.

okuhe: Eleotris fusca (Schneider). This *'o'opu* has more names than any other. On Kauai it is *okuhe* or *akupa*, the young called by either name doubled, *okuhekuhe*, or *akupakupa*. A slight colour variation, with a yellow tinge, is called *okuhe melemele*. On Oahu and Maui, *oau*, or *owau* is the favourite name.

Mrs. Pukui remembers a little story of the *owau* (the pronoun I). Some one caught a lot of *'o'opu*, but the whole catch disappeared. The owner kept calling, "Where are my *'o'opu?*" A voice answered, "*Owau, owau,*" and the fishes turned into lizards and scampered off. Since then, *owau* has been one of the names of this fish.

It is also called *apohā*. Kinney says *okuhekuhe* is the name applied to the young after they have passed the *hinana* stage. Jordan and Evermann (40, p. 479) describe the colour as "dirty brownish throughout, belly paler; fins all dark, soft dorsal narrowly white-edged." Size, according to Mainland (62, pp. 21-25) is up to 6 inches, perhaps up to 9 as maximum. Mrs. Pukui notes that it is the largest *'o'opu*. Scales are small. J. and E. state "very abundant in fresh, brackish and shallow water in the Hawaiian islands." Mainland notes, "Its normal motion is characterized by a slow easy movement along the bottom, poking its head under rocks and bits of vegetation and debris. Occasionally it moves with a fast, jerky burst of speed . . ." It is eaten like other *'o'opu*. This with the *nakea* and *nopili* are the three favourite *'o'opu*.

'O'opu of brackish waters:—

kanī'o. This has bluish stripes, crosswise (Mrs. Pukui); effect resembling the *manini* (Kinney). Not abundant, not much relished as food. It enters streams for a distance, but not the upper reaches (Mrs. Pukui). Mainland identifies it as *Stenogobius genivitattus* (Valenciennes)—(if it is his "*kanī'e nī'e*," for which name I find no corroboration). Colour sandy white, light tan, or olive, with seven to twelve transverse bands of dark brown or black . . . dorsals and caudal light yellow or olive, spotted with brown or black dots . . . Head with a very characteristic broad black band under the eye . . . Size, up to 6 inches, but often not more than 3; rarely found in the fast flowing water of the lower valley (62). Like other *'o'opu* its swift movement is by fast, jerky leaps; it lets itself slowly sink to the bottom after such dashes. It also creeps slowly along the bottom. "When the fish reaches a good feeding ground, it stops and takes sand into its mouth and shoots it out again. When frightened it either buries itself in the sand leaving only its eyes exposed, or dashes under a rock."

kauleloa (this may be Mainland's *pao'o kaulaloa*, which he identifies as *Oxyurichthys papuensis* (Valenciennes). Mainland (62, pp. 58-60) describes this fish as olive grey, trunk and tail crossed by six to eleven bands of dark olive or brown, irregularly spaced and of varying width; five dark olive-brown blotches along the median line, a dark olive spot at the base of the pectoral fins. Olive characterizes the fins, a "dark row of dots at the base" of the caudal fin. Size, 5 to 6 inches. It is most commonly found in brackish water at the mouths of streams. Hawaiians say it likes to hide in holes in the mud;

it was common at Kalia beach, Oahu, some years ago.   It is edible
(scant praise), cooked in *ti* leaves.   (Mrs. Pukui.)

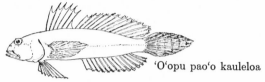

'O'opu pao'o kauleloa

In legend, " The *'o'opu* of Kawainui was famed for not swimming away from the hand of the fisher, but rather clinging to one's
skin in the water."   (25, I: 374.)

'O'opu hue (*makimaki, keke*), swell fishes, puffer fishes, globe
    fishes: *Tetraodon hispidus* Linné and other *Tetraodon*.

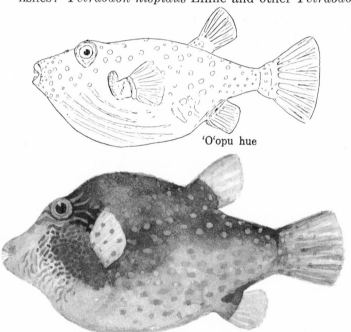

'O'opu hue

Description: The puffers are described by J. and E. (40, p. 424)
as having " body oblong, or elongate, usually little compressed, sometimes very broad, head and snout broad; skin scaleless, usually more
or less prickly . . . jaw . . . forming a sort of beak . . . spinous
dorsal and ventral fins wanting . . .   Fishes of sluggish habits,
inhabiting warm seas, noted for their habit of filling the belly with
air " when disturbed.   They live along the bottom, among seaweeds.
One of the sharp-nosed puffers is called " *pu'u olai* " according to
J. and E. (40, p. 434).   The term means lava flows of the hills, doubtless the term was used because of the colour streaks from eye and
mouth.

The puffers have no relation to the *'o'opu*, and it is not likely
that the Hawaiians thought so.   It is possible that the term is or was

actually *'opu hue* (stomach like a gourd)—perfectly appropriate. The
term *keke* means pot-bellied.   The widely quoted term *makimaki*,
however, is open to criticism (sometimes further corrupted to *muki-
muki* (40, p. 427; plate 67)).   The term *makimaki* is translated
" deadly death " by Jordan and Evermann (40, p. 428).   The term for
death is *make*, not *maki*.   The doubling of *make*—*makemake*—means
something remotely different—desire or wish.   It seems possible that
makimaki is a mistake of some foreigner.   It must be admitted that it
pushed its way into the Hawaiian dictionary.

  T. hispidus is the largest of the Hawaiian puffers, being up to
14 inches in length; olive green with a whitish belly, streaked with
yellow.   (40, pl. 66.)

  It is wisest not to eat this fish at all.   Throughout the tropics
it is known for its poisonous quality, sometimes deadly.   Hawaiians
ate it rarely, thinks Mrs. Pukui.   Malo (63: 73) says: " the poisonous
part is the gall.   By carefully dissecting out the gall-bladder without
allowing the escape of any of its contents, the fish may be eaten with
impunity.   Its flavour is delicious."   Kepelino says (52) :—

 There are two kinds of *'o'ophue*, the *malani* and the
 manalo . . .   The *malani* is a poisonous fish indeed.   The
 'o'opuhue manalo, however, is edible.   To tell them apart look at
 the teeth.   If the teeth are yellow the fish is poisonous.   But if

the teeth are a bright, clear white the *'o'opuhue* is good to eat. The skin is rough and should be peeled off. This fish has much flesh and is good. It should never be eaten raw.

Beckley (4, p. 2) says:—

'O'opu hue is the well-known poison fish of the Pacific, but of a delicious flavour. It is generally speared in enclosed salt water ponds from the stone embankments. The poison of this fish is contained in three little sacs which must be extracted whole and uninjured. The fish is first skinned, as the rough skin is also poisonous in a slight degree. Should the teeth of the fish be yellow then it so highly charged with poison that no part of the flesh is safe even with the most careful preparation. *'O'opu hue* caught in the open sea are always more poisonous than those from fish ponds.

Mrs. Pukui says that one Japanese family of her acquaintance is particularly fond of it. The skin is always peeled off and the internal organs removed, except the liver! The liver is highly esteemed.

Both Chinese and Japanese in Hawaii eat the *'o'opuhue* to a considerable extent, being extremely fond of its delicious flavour. Dr. Nils P. Larsen has written of the danger of eating this fish (56, pp. 417-421), citing the cases of two Chinese who had eaten *'o'opuhue* for years. They ate of one fish, however, and both were dead within four hours, evidently without any warning when eating that this particular fish was poisonous.

Tetraodon hispidus is a fairly common fish in Hawaii, and is found in estuaries most commonly. Other species are pelagic or reef fish, some are rare. (82, labels on fish casts Nos. 202, 203, 205, 206, 207).

O'ouma ma'auwele (wander about): *Stethojulis albovittata* (Bonnaterre). Evidently a fish of the *omaka* type.

Opakapaka (*paka* is the Ka'u name). Hawaiians state that the *opakapaka* are the blue snappers. By this criterion the following fish are *opakapaka*: *Pristipomoides sieboldii* (Bleeker), *Pristipomoides microlepsis* (Bleeker), *Aphareus furcatus* (Lacépède) and *Aphareus rutilans* (Cuvier). Hawaiians named *opakapaka* at four stages of growth: *ukikiki*, under 12 inches, *pakale*, then *opakapaka*, by which name it is best known, and finally *kalekale*, about 2 feet or more. The term *ukikiki* is shared with the red snapper, *'ula'ula*.

Description: *P. sieboldii*, according to cast 480 in Bishop Museum, is about 19 inches long, moderately plump, colour light grey with yellow areas, dorsal, anal and ventrals grey, tail fin dark grey with a wide yellow area in upper lobe, edge of the whole fin yellow. *P. microlepsis* (Bleeker) is not represented in Bishop Museum casts. A description by Jordan, Evermann and Tanaka (42, pp. 649-680,

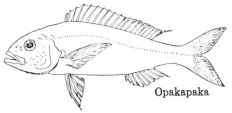

Opakapaka

1927) gives "general colour in life, pale gold and grey blue, inter-mixed over body in fine pattern . . ." Fowler (26: 192-93) gives the colour as "faded uniform brownish, with slight olive cast above, below and on sides lighter or with silvery tints . . ." *A. furcatus*, according to cast 479, in Bishop Museum, is about 24 inches long, body is of a medium dark blue, the greatest depth of colour being at the edges of the scales; pectoral orange, dorsal and anal fins yellow and orange, the spines darkest, tail fin very dark blue or black, the inner edge yellow orange and red. Another cast, 478, smaller, is very like in colour. *A. rutilans*, according to cast 477, in Bishop Museum, is about 20 inches long, colour grey below the lateral line, edges of scales touched with violet, dorsal area above median line purple or violet, scale edges deeper in colour; dorsal fin grey, with top and bottom yellow deepening to orange in soft dorsal; anal grey with posterior spine orange and very long; tail fin and pectorals orange; top of head deep purple, under jaw pale blue, eye ringed with orange.

Eaten raw, dried or cooked in any manner. Flesh is white. This was one of the most common fish on restaurant menus pre-1942. Perhaps it is better liked by foreigners than Hawaiians. Kepelino (52) says, "fleshy but insipid."

A deep sea fish.

'Opelu, mackerel scad: *Decapterus sanctae-helenae* (Cuvier). Varieties are: *'opelu kakala-lei* (small, deep-water variety); *o. kalamoho* (large, plentiful in some areas, notably Kona, Hawaii. It is possible that this is another name for *'opelu palahū*). The young are called *'opelu kikā*, and are delicious when fat.

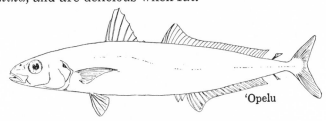

'Opelu

Description: Length, about 12 inches; colour, dark blue, becoming white on belly; upper part of body streaked with deeper colour, fairly straight streaks. Scales fine.

Highly prized as food, eaten raw, dried, sometimes broiled after drying, or broiled when fresh. Kepelino says the dark meat is delicious, and that there are not many bones. A note in the Kelsey collection (51) says, " The *'opelu* is a tender fish that quickly spoils."

The *'opelu* of Waikiki and of Wai'ane were recognized by the plumpness of the body, and the largeness of the eye; delicious.

With the *aku*, this fish figures in one of the best known stories or traditions of Hawaii—the coming of Pa'ao to Hawaii from Kahiki, already related (p. 58 herein).

Vancouver notes, too, in speaking of the *aku* and the *'opelu* (85, vol. III: 17-18) " during those months that the one is permitted to be caught the other is prohibited . . ."

In Ka'u, where there was a great deal of fishing for both these fish, there was a *heiau* (temple) devoted solely to offerings for the abundance of *'opelu*. (Mrs. Pukui.)

'Opelu palahū (*'opelu paka*), chub mackerel: *Scomber japonicus* Houttuyn.

'Opelu palahū

Description: Length, about 16 inches, body slender, not compressed; colour, bluish above, lighter below, silvery, upper body has many crosswise, darker, broken streaks that fade off into spots below lateral line. (40, p. 170.)

Prepared like the other *opelu*.

J. and E. say the *o. palahū* is not as common as *'opelu*. Mrs. Pukui compares the two. The *o. palahū* is three or four inches longer. The *'opelu* has more definite, precise streaks; those of the *o. palahū* are like the daubings of a child with ink-dipped fingers; the lines continue, but brokenly. According to the drawings in J. and E. (40, fig. 62 and pl. 30) the soft dorsal fin differs markedly. All these comments are borne out by casts in Bishop Museum—no. 26: *'opelu;* no. 25: *'opelu palahū*.

Note by Kelsey (51): " This fish is symbolic of a person of unstable character. An over-grown *'opelu* (!) called an *ono*, above which swam a pilot fish, was the leader of the school."

Opule (variegated in colour), one of the wrasse fishes: *Anampses cuvier* (Quoy and Gaimard), *A. evermanni* (Jenkins), and other *Anampses*. Hawaiians name several kinds of *opule: the o. 'ele'ele* (black), *o. lauli* (dark dorsal), *o. lali'i* (fine dorsal), *o. makole* (red-eyed), *o. uli* (dark) and *hilu opule* (variegated).

Description: The *opule lauli*, one of the most common, is described and figured by J. and E. (40, fig. 127, pp. 293-294), as *A. evermanni*, brownish red, with narrow blue line on each scale, head blue or bluish, fins brownish or red, with blue lines and dots; about a foot long, or a little less. The *A. cuvier* is greyish olive, underparts brick red, with

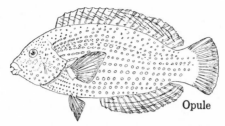

Opule

spots, blotches and streaks of white, and some small yellowish specks; head brownish with pale blue wavy lines; ventral, anal, and dorsal fins blood red with sky blue bars, caudal and pectoral orange (40, p. 292). Scales of all *opule* are large. The *hilu opule* is an *opule*, not a *hilu*, says one informant (Watson). It is a dark brown, with white spots. It reaches a length of about 10 inches. It is fairly common at Mokapu (Oahu); not common in Kane'ohe Bay.

Not eaten raw, good for broiling and baking.

O'u. Listed (59 h).

Nakuina says, "About the size of the *ulua*, 2 feet or so. Green-yellowish in colour, looks like *ulua* but is not *ulua*. It is good to eat cooked but not raw."

O'uku'ukū (small *o'u*), listed (59 h).

Owau, see *'o'opu*.

Pa'a'ahahalalū, see *akule*. (*pa'a'a*—a *tapa* term meaning the small lengths of bark for second quality *tapa*.)

Pa'akahala, see *kahala*.

Pa'akai-helele'i (scattered salt), listed (59 h) and Wetmore (88, p. 95). He claims it is one of the *Hemiramphidae*, and says it is rare.

Pa'apa'a, linked with *Dascyllus trimaculatus* (Rüppell) which would make it another name for *alo'ilo'i*.

Pahaha (swollen), Kalakaua's list (59 a).

Pahau (like a *hau* leaf). List of native fishes (59 f).

Kinney says it is a multicoloured fish, the largest reach about 4 oz. in weight. (Nakuina adds that it is small, flat—that is, compressed?—yellow-greyish in colour, like an old *hau* leaf.) Not abundant, not important as a food fish. Eaten raw, salted or *pulehu*; sometimes added to *maikoiko* and *hinālea* and broiled after being wrapped in *ti* leaves. They live in the coral reef crevices.

Paheuheu. Henriques list (59 d).

Pahikaua (war knife), the same as *kala li'ili'i*. It is also the name of the mussel (Mrs. Wiggin).

Pahu (box) (*moa*, or *moamoa*, or *makukana*), trunkfishes, boxfishes, cowfishes: *Ostracion* species. J. and E. (40, pp. 441-446) list five species for Hawaii; Fowler (26: 461-465) lists seven species. Jordan and Evermann figure and describe the *moamoa wa'a* (40, pp. 443, pl. 51) as *Ostracion oahuensis* J. and E. (This is *Ostracion sebae* Bleeker of Fowler.)

Ostracion have three-, four- or five-sided bodies, and "with distinct teeth in the jaws" (40, p. 441). They are small fishes usually. The *mahukana* of J. and E. (40, p. 445) is Fowler's *Ostracion fornasini* Bianconi. It has a pair of spines projecting forward above the eyes, one on each side of the body, another forward of the anal fin, and one high on the back. Nakuina says these fish look like coffins, are dark grey or blackish in colour, and are finger length or so. Fowler (26, p. 9) says "bottom dwellers in shallow water." There is little flesh on them and they were not eaten. They are included here because they are said to be poisonous in some parts of the Pacific (89, p. 9) and they may be in Hawaii.

Paka, the Ka'u name for *opakapaka*.

Pakaikawale, a variety of *opakapaka*? Listed (14: 498).

Pakala, pakalaka, pakalakala, see *kala*.

Pakale, see *opakapaka*.

Pakaueloa, another name for *olali*, the young of the *hou*.

Paki'i (flat), flounders: *Platophrys pantherinus* (Rüppell), *P. thompsoni* (Fowler), *P. mancus* (Broussonet), *Engyprosopon hawaiiensis* (Jordan and Evermann), and other flounders. Only two Hawaiian names have been received, *p. moana* (ocean), and *p. a-ha'awali* (*ha'a*—dance in a crouched position; *wali*—soft and graceful). *Bothus* is a synonym of *Platophrys*.

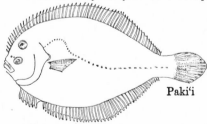

Paki'i

Description: Length of most varieties is from 8 to 10 inches (in Kane'ohe Bay, Watson). Nakuina says they reach 3 feet or more. Colour, always darker on the upper side than the underside, which is distinctly white. The upper side is dark brown or light sand colour, mottled or spotted. This fish is well camouflaged against a background of sand, and when danger is near it clouds the water near by with the sand stirred up by burying itself so that eyes alone protrude. Takes on the colouring of its surroundings (Nakuina). Scales are small.

The flesh is dainty, white in colour, it is usually broiled, sometimes only partly dried, then broiled. A delicious fish; not eaten raw.

Very common; often caught by jabbing a sharp-pointed stick into the sand while walking through the shallows of the reef waters. If the stick quivers, a *paki'i* may have been speared.

Pakiki, see *paki'i*.

Pakole (like the *kole*), listed (59 i).

Paku'iku'i (patched), one of the surgeon fishes: *Acanthurus achilles* (Shaw).

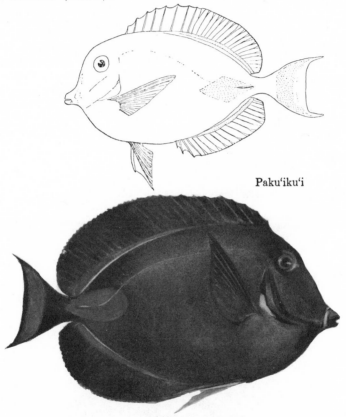

Paku'iku'i

Description: About 8 inches long; 6 is about the maximum in Kane'ohe. Colour varies from light to dark brown with touches of orange and pale blue at bases of dorsal and anal fins; lozenge-shaped orange spot at caudal spine; touches of light blue on ventral fins and on head. Scales very small, caudal spine depressible in a groove.

Good to eat, always cooked, excellent broiled.

Abundant; one spot where they were noticed was in a little cave of an island off the windward side of Oahu where the fish remained in the foam of each breaking wave, swirling to face the next wave when the waters cleared a little.

Palae'a, see *olani,* the young of the *hou.*

Palaelae, young of the *lai,* or *lae.*

Palahoana: *Brotula multibarbata* (Schlegel).   Hawaiians
   sometimes called this *puhi* (eel) *palahoana.*

Palahoana

Description: About 12 inches long; colour, "raw umber, paler
toward belly, fins nearly black along outer portion" (40, p. 507).
Cast 142 in Bishop Museum is about 15 inches long, mottled brown,
with the outer edge of dorsal and pectorals orange.   Body is plump,
thinning out to an eel-like tail.   Scales small.

Good to eat cooked, not raw.

This fish stays in pools at water's edge;   not in deep water;
used to be plentiful in Hilo (Nakuina);   very few at Kane'ohe
(Watson).

Palaila.   List (75.37).

Palaka.   Listed (59 d).

Palani, one of the surgeon fishes: *Acanthurus dussumieri*
   Cuvier and Valenciennes. Some say that *maiko,* or
   maikoiko is the name of the young; others say *maiko*
   is a distinct fish. See *maiko.*

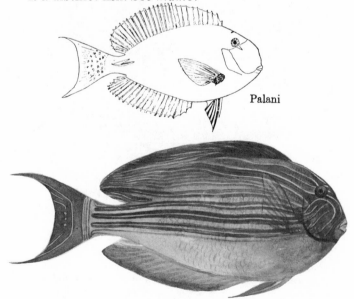

Palani

Description: One informant says about 6 inches, another that the species (more than one?) vary from about 6 to 12 inches in length; colour, brown, says one informant; J. and E. (40, p. 391) describe the fish as dull bluish olive, with brassy and bluish markings and shades, a blue line along base of dorsal fin; caudal fin bluish with blackish olive spots. One informant says *palani* is orange, but the fins are more brownish than the body. There seems to be disagreement among Hawaiians as to which surgeon fish is the *palani*, which the *puwalu*. Mrs. Pukui and Mrs. Leong say that of the two the *palani* is darker in colour and stronger in odour.

As food, well liked in spite of the strong odour of skin and flesh; skin always removed; eaten raw, or broiled, or cooked in a calabash. According to Mrs. Pukui there is a Hawaiian riddle: "*Ku'u i'a pa i ka lani*" (My fish whose odour reaches heaven).

The mythical origin of the odour of the *palani* is given in a tale of Ke'emalu (75.24) :—

> As she floated around in the ocean she recalled what her grandmother, Hina, had told her, that she had an ancestor among the fishes of the sea, named Palani-nui-mahao'o. She called to him and in a short time she found herself on her ancestor's back, being borne shoreward. As she was taken back to shore, she was seized with such a desire to urinate that she was unable to control herself and so she urinated on her ancestor. Her ancestor became very angry and left her out at sea. It is said that was how the *palani* got its strong odour. When she found herself deserted she chanted a chant of derision to this ancestor . . .

Another legendary account of the origin of the odour, given in the tale of Punia (75.47) is that the hero, Punia, killed the ghosts, multitudes of them and rolled them up in a fish net. The ghosts must have tainted the nets, for this seems to be sufficient reason for the story without further explanation.

For the method of getting rid of the odour of the *palani*, see directions under the *kala* fish. Broiling was the best method of removing the remaining odour. Mrs. Pukui records another belief. "There are some people who have an especially strong odour of perspiration—very unpleasant, but it makes them immune from skin diseases. This condition is called *'ili-'awa* (poison skin). The odour of a strong-smelling fish is said to be intensified if it is caught by an *'ili-'awa*, even noticeable by the man himself. If the odour has not been removed from the *palani*, it clings to the breath of those eating it as tenaciously as does the odour of onion."

According to a brief note (75.39), this fish was *kapu* to men, but free to women.

Paloa, listed (59 d).

Palaoa: the sperm whale, or any whale.

Not sought for food, not eaten. If a whale became stranded, or drifted ashore, it was the property of the chief of the district. The teeth were greatly prized as ornaments, and the ornament carved from a whale's tooth and hung from braided strands of human hair was called *lei palaoa*.

Palaolao, listed (59 d).

Palapala (marked), the Maui name for *pualu*.

Palau, see *hinālea*.

Palaulau, listed (59 d). A species of red fish.

Palemo, the young of the *uhu 'ula*.

Paloa, listed (59 g). Possibly a misspelling.

Palukaluka (slimy), a variety of *uhu*.

Pamomoa, listed (59 a).

Panohonoho, the young of the *nenue*.

Paniholoa (long-toothed) ; one of the wrasse fishes: *Thalassoma trilobata* (Lacépède).

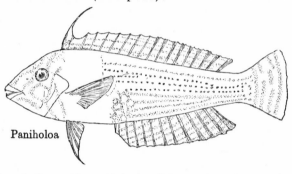

Paniholoa

Description: About 18 inches long; colour, brilliantly coloured, body rosy red with two rows of green rectangular spots from pectoral to caudal fin; dorsal, anal and caudal fins yellow, edged with blue; head and breast yellow (described from 40, pl. 34).

Kanakaole says, " a delicious fish, fairly common."

Panuhu, or *panuhunuhu*, see *uhu*.

Pao'o, see under *'o'opu*.

Pa'opa'o, see *ulua pa'opa'o*.

Paouou, or *pahouhou*, see *hou*.

Papa-hapu'u, listed (59 h, 8/26/99). Possibly a *hapu'u*.

Papai, see *po'ou*.

Papiopio, or *papio ulua*, see *ulua*.

Pauhuuhu, *uhu-palukaluka*.

Pa'upa'u, or *pau'u*, or *pau u'u*, see *ulua*.

Pihā (or *pihā-ā*): *Stolephorus delicatulus* (Bennett).

Kamakau says "Waihe'e, Maui, was a land where fish were abundant . . . The fish that came in schools were the *nehu* and the *pihā-ā*." (75.20.) Mrs. Leong says they are "very similar to *nehu*, a little larger, perhaps a large *nehu*. Sometimes the sea, up to the very shore is white with them when a school comes in to the shore. The largest is about 3 inches long, and they are all silvery. They were seldom eaten raw, but were best dried and salted."

Pihāweuweu (full of wild grasses and other low weeds). Listed (59 d).

Piliko'a (coral clinging) : *Paracirrhites cinctus* (Günther), *P. arcatus* (Cuvier), *P. forsteri* (Schneider), the last referred to as *hilu piliko'a* (evidently *piliko'a* of the *hilu* type). *Piliko'a*, besides being used to designate this fish is used as a descriptive term for some other fish, as the *ala'ihi piliko'a*. The name *hilu piliko'a* is unusual in that the fish is a *piliko'a*, not a *hilu*, and the descriptive term usually has second place. (Compare, however, *hilu opule*.) In J. and E. (40, p. 449), *po'opa'a* is given as an alternate name for *P. cinctus*, but Hawaiians deny this statement. *P. cinctus* is also referred to as *'o'opu kahaihai*, or *'o'opu piliko'a*, and how widely the *piliko'a* is so called, I do not know.

Piliko'a

Description: Length, 8 inches or less; colour, light pink, in general. J. and E. (40, p. 449) describe the three above named as *P. cinctus* as having four broad pink bands with white between, from dorsal to ventral, head dusky green, fins lighter green, except for a red dorsal. *P. arcatus* is olivaceous with fine red stripes from head to tail, fins greenish; it is 2¾ to 5½ inches in length (40, p. 450). *P. forsteri*: head olivaceous, speckled with bright red, body has yellow stripes lengthwise, with broad area of black from middle body to and into the tail, fins reddish, except anal, which is yellow; 4 to 8 inches long (40, p. 450). Watson says, " Closely related to *po'opa'a*, scales hard, large, difficult to scale; common inside the reef, fins and tail bright red; in Kane'ohe it is usually thrown away by fishermen."

Pilipohaku (clinging to the shore). Wetmore list (88).

Pinao (dragonfly) ; see *lolo-au*.

Poalani, listed (59 a).

Poe (rounded), listed (59 h, 8/26/99).

Poapoai (circles), listed (59 d).

Pohopoho (patches), see *kawakawa*.

Ponuhunuhu, see *uhu*.

Po'olā, see *'ama'ama*.

Po'opa'a (hard head) ; *Cirrhitus pinnulatus* (Schneider) ; *Merinthe macrocephala* (Sauvage) ; and *Sebastapistes asperella* (Bennett) ; probably others.

Description: *C. pinnulatus*: body short and stout, 4.4 to 9.75 inches long, moderately compressed, head heavy. Scales large, nape,

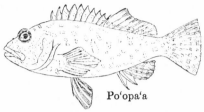

Poʻopaʻa

opercle and breast with large scales, cheeks with very small scales; colour, body marbled and blotched, with bluish, olivaceous brownish and white, with numerous red spots of varying sizes . . . white appears as five ill-defined vertical bars; head bluish white. One of the largest and most important species of the family . . . abundant in the Hawaiian islands (40, p. 452). *Merinthe macrocephala*: head olive brown, finely vermiculated and spotted with bluish and greenish white changing to red and white on lower part of cheek and jaw, side with broad white cross bars alternately with broader red bars. Blotches, streaks and dots of one colour laid over or next to another; the general effect is grey or olive green ground with red and white mottlings. A gayly beautiful fish, used to be common in the market, scales firm, moderate, none on top of head or snout. Length about 4 to 5 inches (40, p. 449).

Opinions differ as to value of the *poʻopaʻa*. One expression is: " The fisherman who fools around in shallow waters takes home a poʻopaʻa fish " (*Hokai ua lawaiʻa o ka kai papaʻa, he poʻopaʻa ka iʻa e hoʻi ai*). Some say not much esteemed as food, though it is tasty and flesh is not bony. Some, however, claim that it is delicious, full of meat and with good keeping quality. It is eaten raw, broiled, or salted and dried. When wanted after salting and drying, the salt is rinsed off, and the fish heated, a little water added just before serving.

Poʻou (papai), Cheilinus unifasciatus (Streets); *C. bimaculatus* (Valenciennes); *C. trilobatus* (Lacépède), perhaps others. Hawaiians distinguish between one that was whitish, *poʻou kea,* and one that was distinctively more red, *poʻou ula.* (John E. Randall considers that *C. rhodochrous* is preferred name.)

Description: All are " usually bright coloured, the shades chiefly red and green " (40, 319). *C. unifasciatus*: 5 to 10 inches long (other species are about 5 inches long); colour, head livid violet brown,

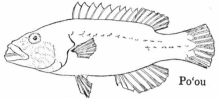
Po'ou

orange and scarlet marking, body reddish brown, each scale with a bright red bar, caudal peduncle olivaceous with a light pinkish bar, tail fin has blue, the tip red; dorsal dull olive green or brown, with orange marks or lines; anal and ventrals rosy, with a red and black blotch on inner side, pectorals orange, scarlet at base, jaws and throat livid bluish (bones and teeth bluish green); another example considerably different in colouring; other species equally gay, brilliant, many-hued (40, pp. 319-322).

Eaten raw, broiled, every way; a good fish, better than *hinālea*, tender, soft flesh (Watson inf.). A fish found daily in the fish markets of Hilo and Honolulu in 1901 (40, pp. 319-322).

Caught all over Kane'ohe Bay (Watson).

Popolo (also the name of a *Solanum*, a common Hawaiian weed with purple berries); listed (59 a).

Poupou (stocky of figure), see *kupou*.

Pua. The term means flower, and also the young of many fishes; note such names as *puahole*, young of the *aholehole*, *pua 'ama'ama*, young of the *'ama'ama*, also *pua humuhumu*, *puakahala*, *pua 'o'io*, *puawa*, *pua uouoa*.

Puai'i, the Andrews-Parker dictionary (2) says "the fry of the mullet" (*'ama'ama*).

Puaki, see *puhi*.

Pualu (*puwalu; palapala* on Maui), one of the surgeon fish: *Acanthurus xanthopterus* Cuvier and Valenciennes. This fish closely resembles the *palani*, and the young of both fish are called *maiko*, or *maikoiko* by some.

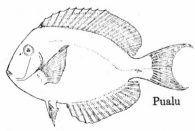
Pualu

Description: Up to about 8 inches in length; colour, generally brown or dull grey. According to Kanakaole, the distinguishing mark of a *puwalu* among the surgeon fishes is the blue line across the soft part of the tail fin.

Like the *palani*, it has a tough skin, and a strong smell, but nevertheless is relished by some, raw or broiled. Broiling seems to remove some of the disagreeable odour. It is good for *i'a ho'ohauna*, and is also used dried.

Puhi, eels. There are numerous eels in Hawaii. Jordan and Evermann (40, pp. 73-113) recorded six conger eels, one snipe eel, eight snake eels, one of the family *Moringuidae*, thirty-five morays.

The following list of Hawaiian names of eels is doubtless incomplete; some names may be local variations for names already in the list.

puhi hoana, not an eel, see *palahoana*.

puhi ho'ola'au (stick-like).

puhi 'ini'iniki (pinch), a small black eel, about the size of a lead pencil, edible.

puhi kāpā (fierce): *Echidna nebulosa* (Ahl). This is a fierce, voracious eel, a fighter. It is about 30 inches long, " the dark areas are dark brown with chrome yellow spots, the bars between these dark areas grey and brown, anterior tubular nostril orange; iris orange." Another example varied somewhat from this colouring. " The natives say that this eel goes ashore in the grass wriggling quickly to the water again when disturbed. They also claim that it is savage and will bite " (40, p. 111). It is said to climb *hala* trees along the shore and has been known to drop down on persons below. It moves swiftly over wet rocks, but cannot move through dry sand.

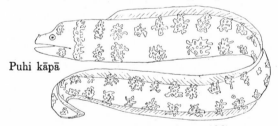

Puhi kāpā

puhi kapa'a: *Gymnothorax picta* (Ahl).

puhi kapipi (salt lightly).

puhi kauila: (name of a tree, wood is dark brown): *Muraenophis pardalis* (Schlegel). A large, dark brown eel. Same as *puhi oa?*

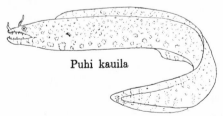

Puhi kauila

puhi kina'u, a small white eel, used in medicine; also eaten as food.

puhi kowali (morning glory), very white, with stripes around the bottom.

puhi kumuone (sandstone), sand coloured.

puhi laʻau (wood), *Myrichthys maculosus* (Cuvier).

puhi laumilo (*milo* leaf), *Gymnothorax undulata* (Lacépède).
J. and E. record *laumili* (probably an error in spelling)—the same?
(40, p. 98). Rare, highly relished. In the " Song for Keawenuiaumi "
(25, vol. III: 467) there is a line, " The yellow colouring of the
laumilo eel, *kapa*-like."

Puhi laumilo

puhi leihala (*hala lei*), *Echidna polyzona* (Richardson), yellow
and brown ring around the body (doubtless suggesting a *lei* or neck-
lace of *hala* or pandanus fruits, one of the most popular *leis* of olden
days).

puhi nanaka (a break that shows another substance underneath),
a black and white eel, will attack and bite.

puhi nauʻai (food chewer).

puhi nonokea (very white).

puhi nukuʻula (red mouth).

puhi ao, *Muraenophis pardalis* (Schlegel), therefore the same as
kauila.

puhi kuna, a fresh water eel.

puhi omole (smooth all over), a white eel.

puhi oniʻo (spotted), a spotted eel.

puhi opule (variegated in colour), a short, spotted eel.

puhi ou (protruding), *Gymnothorax petelli* (Bleeker), has red
and white speckles.

puhi paka (fierce). One of the best liked; found throughout
Hawaii; teeth very sharp, a ferocious eel. " They remain in their
haunts at night and venture out when enticed by bait (*hauna*):
pounded heads and entrails of any fishes. When lured out in this
way they will catch at hooks with any kind of bait, especially when
coming up a swift current heavily baited." (Kinney). One writer
from Lanai (75.26) says, " We have some four or five species of eels
and one is a terrible fellow . . . found in rocky shoals, and in the
coral ponds isolated by low tide. He often baffles the efforts of the
fisherman. He will swallow the hook and bite the line in two. He
will force himself out of a net, and if you have got him with a stout
hook and line you must tear him to pieces before you can drag him
out of the hole in the rocks in which he has set and braced himself.
But occasionally he is found away from his rocky fastnesses and then
our Hawaiian Neptune has a chance at him with his trident and he
seldom misses a stroke. This dangerous eel, the *puhi paka*, inspired
terror among the Hawaiians. He will take off a toe, or snap off an
exposed naked foot, if he gets a chance. Where he is found no crabs
or little fishes are to be seen in the pools nearby. He devours everything."

puhi palina (having cheeks).

  „   *papaʻa* (burnt), same as *puhi kapaʻa?*

  „   *pule* (variegated in colour), small, spotted.

  „   *uhā, Conger cinereus* Rüppell, ventral side white, dorsal darker, a fathom in length.

puhi uha kalakoa (*kalakoa*: calico, modern word).

  „   *ula*, same as *puhi uhā?*

  „   *wela*, another name for *papaʻa*, or *pakaʻa?*

The young of all eels, from about 1 foot to 2½ feet in length are called *ʻauau ki* (*ti** stem), or *puaki* (*ti* blossom), or *oilo*. Above this size they take their distinctive names.

Salt water eels were eaten, but freshwater eels were used ceremonially only. Of details of this use there is no available record. All eels required cleaning. Entrails were removed and the eels put into a vessel with a handful of salt, and worked around with the salt until all slime was loosened from the skin. Then the slime was washed off and the process repeated until it was entirely removed. A cut was made on each side of the spinal column, and the vertebrae removed. The flesh was then salted and dried. When wanted, they were ready to cook over the hot coals or in the *imu*, wrapped in *ti* leaves. Eels were not eaten raw.

Small eels (*puhi oilo*), about as big around as a finger, were wrapped in *ti* leaves and cooked until half done. Then the wrapper was opened, the meat removed from the spine and head by holding the eel by the head and pulling off the flesh. The flesh was seasoned with salt, rewrapped and cooked until done. (Wiggin.)

Some eels have a strong odour, unpleasant. One informant recalls seeing wood ashes used to remove the slime and reduce the odour of these eels. (Charles Alona.)

Eels were highly prized as food. " The eel was a fish of which chiefs were fond . . . so much prized by those of Koolau, Maui . . . that they said only beloved guests were served with eels . . . for eels were considered choicer than wives . . . When we arrived we were served with eels of the *uhā* variety, well dried and broiled over the coals . . ." (75.11.)

Eels were caught in various ways: 1. Headfirst in a scoop net, then running quickly to the dry beach to kill it and put it in a bag;

2. By basket traps. Kamakau speaks of this method (75.21):

> In the morning the basket traps were brought filled with eels. The baskets were carried ashore and the eels poured out. With one blow of the mallet on the tail and one on the head, each was killed. Then the *imus* were lighted, the meat wrapped in *ti* leaves and cooked until well done. Then the men, women and children ate heartily of the eels until they were all consumed. The eels caught in basket traps were the eels that were much eaten.

3. By spearing; 4. By hook and line; 5. By " eel pinching " (*ʻiniʻiniki puhi*). Mrs. Pukui gives the following description:

* *Ti: Cordyline terminalis* (Kunth).

Every fisherman knows where the baby eels are to be found in holes among the rocks or, in the sand. The bait is baby squid (*he'e pali*). The squid is held in the palm of the left hand with the tentacles hanging between the fingers, and the hand is held over the spot where the eels gather, resting on the sand or on a stone so as to keep the hand perfectly still. The eels smell the squid and come out to seize a tentacle. With the right hand the squid is drawn back toward the wrist until the heads of the eels appear between the fingers. Then the fisherman closes his fist quickly so as to catch the eels, two to four at a time. If larger eels appear the fisherman leaves for other grounds, for a large eel would mangle his hand if so caught.

Puhiki'i, see *mālolo.*

Puihi, see *iheihe.*

Puili (gathered together). Listed (55 h). Kinney says it is related to the *iheihe,* and *aha.* It is more roundish, less flat-bellied than the *iheihe.* It is most abundant along the Waimea-Kekaha shore of Kauai. Cobb records it on his list (14, p. 441), as "*pu'uili,* half-beak."

Pukukui (crouching tightly from cold). Listed (88).

Pula (irritation in eye, also a leafy branch used to prod into holes in reef to drive fishes into a net): *Pempheris mangula* (Cuvier) (?) (55 e). (Coloured illustration of *P. mangula* non-Valenciennes, 32, pl. 51 B, p. 102.)

Pu'u (heap), see *eheula.*

Pu'u ola'i (volcanic cone, literally earthquake hill). This name is recorded by J. and E. (40, p. 494) alone, as a sharp-nosed puffer: *Canthigaster rivulatus* (Schlegel). Puffers are discussed under '*o'opu hue.*

Puwalu, see *pualu.*

Uhā (fat-cheeked), see *puhi uhā.*

Uhu (persistent in going ahead wilfully), parrot fishes: *Scarus* species, as *Scarus ahula* (Jenkins), *S. perspicillatus* (Steindachner), *S. rubroviolaceus* (Bleeker). Hawaiian names are: *uhu 'ahu'ula* (feather cape), *u. uliuli* (dark), *u. palukaluka* (slippery), *u. 'a'a* (daring), *u. 'ele'ele* (black), *u. halahala* (dissatisfied), *u. la'uli* (dark dorsal), *u. maka'ika'i* (travelling), *u. ahi'uhi'u* (timid), *u. panoa* (desert-like), *u. 'ula* (red) (probably the same as *u. 'ahu'ula*), *u. kualakai* (somersault in sea). Stages of growth are: *'ōhua,* spawn; *ponuhunuhu,* or *panuhunuhu; uhu.* Another informant gives, *male,* or *omamale* as the name of the young.

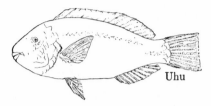

Uhu

Description: In all *uhu* the body is compressed, *uhu* are plant-eaters, the teeth are strong and beak-like, well fitted for clipping off food from coral; scales are large. *Uhu 'ahu'ula*, or *u. 'ula*: *S. ahula* (Jenkins), the red parrot fish, size, up to 36 inches; colour, dull red, except for a violet line on outer margin of dorsal and anal fins; another specimen is "brown, washed with red" (40, p. 347) said to be "not very common (?), and brings an extravagant price in the markets, being eaten raw at native feasts." Scales large, not present on head. *Uhu uliuli*: *S. perspicillatus* (Steindachner), one of the largest *uhu*, about 3 feet; J. and E. note (40: 348) that the colour is bright green, scales on sides edged with brown, belly lavender, head brilliantly coloured with lavender, sky blue, dark blue, brown, light green, a broad rectangular band midway between snout and eyes; dorsal and anal fins striped with green and brown, with blue edge. Watson says big ones weigh 18 to 20 pounds. *Uhu palukaluka*, or *u. kualakai* (in Kaneohe) *S. rubroviolaceous*, size, 7 to 12 inches; colour, reddish brown above, underbody bright red (40: 353). Watson says "more black in colouring of scales." Notes of the other Hawaiian names are: the *uhu 'a'a* likes to fight other *uhu*; the *uhu 'ele'ele* is green; *uhu maka'ika'i*, is probably a descriptive term of a habit of all *uhu*, that is, travelling along one after the other, in line; *uhu ahiuhiu* (timid) is blue, such as those found at Kawaihae, Hawaii; *uhu panoa*, is probably a rare species, the name meaning not numerous.

Uhu are favourite fish with the Hawaiians, sometimes eaten dried or broiled, but usually eaten raw. The flesh is soft, white and a little mushy. It is preferred when combined with pieces of the fat liver. Kamakau mentions this (47):

> When the fishermen's meal was ready there were sour poi, balls of sweet potatoes, pieces of *uhu* fish mashed with fat chunks of *uhu* liver, mixed with finely pounded *lipa'akai* seaweed, a dish of *uhu* fish cooked with hot stones in a calabash, with gravy *heavy* with the fat of the liver, *uhu* cooked with pieces of liver inserted in the flesh, and the cups of *'awa*. They ate heartily, ate till they were filled, ate till they could eat no more.

Another says: "The *uhu* has the most delicious liver. The meat is not as delicious or as fat as the liver, and if the liver is very fat the combination of the two is the best." (75.15). Another says, "The spawn has a deliciousness all its own." (75.28). Beckley (4, p. 15) speaks of *uhu*, rock fish, "two species, red and green. The red . . . are the more choice for eating raw. The green, called *uhu 'ele'ele*, are not so fine flavoured, but attain a large size."

For fishermen in ancient days, the *uhu* was the most telltale of all fish, they revealed what sort of behaviour was going on at the fisherman's home. If the *uhu* capered and frolicked in the water it was a sure sign of too much levity at home, instead of the sober conduct a fisherman's wife should display when her husband was at sea. If two *uhu* seemed to be rubbing noses, it was a sure sign that there was flirting going on at home. The only course open to the fisherman was to quit fishing and go home and punish his wife. (Mrs. Pukui.)

Uhu are found along all shores of Hawaii, and travel in schools. There is always a leader in a school, and they move along, sometimes in single file, sometimes in double file, after the leader. The term for this formation is *uhu-holo*, or *uhu-maka'ika'i*. Makapu'u, Oahu, is a favourite place of *uhu*. When catching by hook and line, the bait used was the *'ala'ala* (ink-bag) of the octopus (*he'e*). It was rubbed over the hook and sufficient smell remained to attract the *uhu*. Every hook let down was apt to catch a fish, but if a miss was made, merely injuring the fish, not catching it, the fishing was over for the day, no more would bite. A special kind of trap was built for *uhu*. Watson describes one built by his own parents in Kane'ohe Bay. It was 8 feet square, built in a spot where the depth of the water was from 18 inches to about 5 feet. It had two gates, and was built at a channel in the reef where the fish habitually file through, called a *ku'una*. When the season came—May, June and July—the outer gate was opened, allowing the leader to come in with his followers. The gate was then shut and the other gate opened as soon as enough *uhu* had been taken for use. Many schools of *uhu* come during a season, so this trap could be used many times. The trap was called *ahu*, the gate *ohi'a*. At Mokapu, at one end of Kane'ohe Bay, is a spot called Keawanui, where *uhu* used to come to feed, and Hawaiians used to keep it in order by eliminating seaweeds inedible to *uhu*.

The *uhu* has a prominent place in legend in the tale of Puniakaia (25, vol. II: 156-163).

Ushering in the tale is an incident featuring Uhumaka'ika'i, a supernatural fish. This fish is a character in part of the story, though he speaks no lines. An abstract of this incident is as follows: While Puniakaia was living with his parents, a desire to go fishing came upon him, so he accompanied his mother to the beach and they went fishing. The kind of fish caught was called *pauhuuhu* (the young of the *uhu*), but only one. The fish was brought home by Puniakaia and cared for and it grew to be a very large fish, so it was given the name of Uhumaka'ika'i (roving, sight-seeing *uhu*). " This fish was the parent of all fishes " (!) Punia turned it back into the ocean when it was full grown, and later, when there was a call for everyone to go fishing, Puniakaia called upon his pet fish in the following manner:—

> Say, Uhumaka'ika'i,
> Crawl this way, crawl this way,
> Draw along this way, draw along this way,
> For here am I, Puniakai'a.
> Send the fish in large numbers
> Until the beach here is stenched;
> The pigs will eat until they reject them,
> And the dogs will eat until they waste them.

Uhumakaikai obeyed instructions, the supply of fish " reached from way down deep in the sea to the surface, and they were driven clear up onto the sand . . .   People began taking up the fish; some were salted, some given away to the people from Makapuʻu Point to the point of Kaʻoiʻo, at Kualoa.   There were still a large number left and the pigs and dogs ate of them . . ."

Kawelo, another legendary hero, was the one to kill this supernatural fish (76: 55-56).   He went fishing, taking along with him a celebrated fisherman, Makuakeke.   The story continues: Kawelo insists on staying out very late, and paddling his canoe to the very spot, off Waianae, Ohau, where the fish are apt to be.   Uhumakaʻikaʻi does not appear, but early the next day, Kawelo tries again to find the fish.   " Soon Makuakeke saw the storm clouds gathering in the sky and knew the fish god was coming.   As the huge fish swam in the sky towards them Kawelo threw his net and caught him.   Then the fish, pulling the canoe with him, swam out to sea until the men could no longer see their homes or the surf beating on the shore.   They went so rapidly that they soon came to Kauai, where the fish turned and swam back with them to Waikiki.   There at last the men were able to kill him."

ʻUiʻui (squeak), see *humuhumu.*

Ukikiki, see *opakapaka.*

Ukole (raw looking), see *kole.*

Uku, a snapper: *Aprion virescens* Valenciennes.   A descriptive term or a variety is *uku palu.*

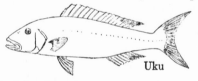

Uku

Description: Length, up to 24 inches (40: 240) ; Watson says up to 3 feet; colour, light grey, upper parts tinged with blue, becoming a very dark blue on top of head; underbody lighter.

" One of the best food fishes," say J. and E. (40: 240), common at the market in 1901.   One Hawaiian referred to the *uku* as the " fish of Kahoolawe," one of the smaller Hawaiian Islands, near Maui, where there are deep waters up to the shore.   Kamakau and Watson note that it is a deep sea fish.   A cast (482) in Bishop Museum measures about 25 inches, and is coloured slate grey all over, with faint green-yellow tinges overlaid.   Dorsal is a lighter grey, spines whitish, membranes darker, three black splotches between the 6th and 9th spines; other fins light grey except tail fin which is very dark grey; pectoral has iridescent touches.

ʻUla lau au. This names appears in J. and E. (40, p. 231) for a fish closely related to the *alalauwā,* and it seems likely that the name is an error in spelling, as there is no fish named *ʻula* in Hawaii.

ʻUlae (*kule*), lizard fish: *Saurida gracilis* (Quoy and Gai-

mard) and *Synodus japonicus* (Houttuyn). Hawaiian
names of species are *'ulae 'ula* (probably *S. japonicus,
'ula* meaning red, and this fish being the more reddish
of the two) and *'ulae niho-a* (gleaming tooth). (John
E. Randall denies *S. japonicus* as a Hawaiian species,
and identifies *S. variegatus* as most common for
Hawaii.)

'Ulae

Description: *S. japonicus* about 10 inches long; colouring varies,
the ground colour white or light grey or a rose red, markings reddish
brown, or olive, or green, darkest on the back, sometimes a faint bluish
band running through the other colouring, belly white or whitish. In
some cases the dark mottlings are of the deepest scarlet, others brick
red, while those found on sandy shores are olive green. The species
is found in two colours, red and green, on the coasts of Japan, as well
as in Hawaii. (40, pp. 63-64.) A cast of this fish in Bishop Museum
(No. 120) is about 12 inches long, the colouring as follows: basic
colour a dirty white, with mottlings of light brown, verging toward
red on chest and head, seven rather distinct, large, mottled splotches
of reddish brown, centring along the median line; fins whitish, barred
with tawny yellow; eye blue, encircled with red. *S. gracilis* (cast 121)
is about 10 inches long, basic colour tawny yellow, mottled markings
greyish brown. Bodies of both fish are roundish, *S. gracilis* (cast 121)
more slender, head more pointed. Both have large scales and sharp
teeth. Both are common reef fishes. These fishes are usually broiled,
with or without a wrapping of *ti* leaves. The name *'ulae* has been
given to the freshwater carp introduced from the Orient. (Simon
Nawaa, inf.)

'Ulae mahimahi, see *maka-ā.*

'Ula'ula (ma'ula'ula) (red), the red snapper. Hawaiians
state that red snappers are *'ula'ula,* and by this criterion
of colour the following are therefore assumed to be
'ula'ula: Apsilus zonatus (Valenciennes), *Etelis
marshi* (Jenkins), and *Etelis carbunculus* (Jordan and
Evermann. Hawaiians speak of the *'ula'ula* (dark *'ula-
'ula*); *u. koa'e* (resembling the *koa'e,* or tropic bird),
and *u. maoli* (native, indigenous, common). The young
stage of all *'ula'ula* is termed *ko'i,* another term *ukikiki.*

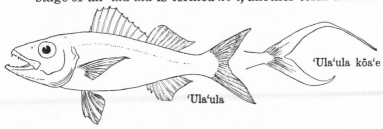

'Ula'ula kōa'e

'Ula'ula

Description: *A. zonatus*, according to a cast in Bishop Museum (No. 481), is about 17 inches long, coloured similarly to J. and E. (40, pl. 16). They note (40, p. 233): "upper half of body with four broad yellow bands, last one extending to base of caudal, between these are three light red bands, nearly as broad as the yellow ones; lower half of body yellow, the edge of scales here tinges with red; head and snout bright golden red . . . dorsal bright yellow . . . margin of soft dorsal tipped with reddish; caudal bright yellow, with reddish tinge, end yellow . . . anal membrane faint golden red, rays faint red; ventrals pale, tinged with red; pectoral membrane pale, rays light yellow . . . scales rather small, firm . . ." *E. marshi* is represented by two casts in Bishop Museum, one (479) about 21 inches long, the other (368) about 30 inches long. J. and E. (40, p. 241) describe this fish as "rose-red, not quite so brilliant as *E. evurus* (Fowler's *E. carbunculus*); a golden stripe along lateral line, mouth not red inside . . . belly silvery, but less abruptly so than in *E. evurus* and somewhat shaded with rose; fins rose-coloured, the first dorsal and caudal brightest, ventrals and anal almost white . . . scales moderately large, firm, deeper than long . . ." In Bishop Museum casts the body is very plump, tail fin alone is bright red, other fins yellowish with rosy touches. *E. varbunculus*, which can be definitely linked with *'ula'ula koa'e*, for according to Mrs. Pukui, it was so named because of the bend of the upper lobe of the tail, resembling that of the wing of the tropic bird (*koa'e*) in flight, is represented by cast 475 in Bishop Museum, about 38 inches long. J. and E. describe this fish (40, p. 242) as "rather long, tapering, moderately compressed; dorsal outline slightly convex, ventral outline nearly straight . . . eye very large . . . colour . . . brilliant rose-red, the side from level of eye abruptly silver, with rosy shades; snout, jaws, eye and inside of mouth red; fins all rose colour, the dorsal and caudal bright; ventrals and anal pale, the former washed with red on centre . . . pectoral pale rosy . . . scales moderate, firm . . ."; a second cast is about 36 inches long.

In a catch of eleven, in deep water, the size varied from 14 to 26 inches.

The flesh of all these snappers is white, there are few bones, the statement is that it is delicious raw, dried or broiled.

It was sometimes used in sacrifice when a red fish was required.

Ulua, certain species of *Carangidae*, crevally. Stages of growth are *papiopio* (now sometimes shortened to *papio*), *pau u'u*, or *pau'u*, and *ulua*.

There are several species, the two largest and most common being *ulua aukea* (light): *C. ignobilis* (Forskal), called the white *ulua*. These both reach a length of about 5 feet and a weight of 100 to 125 pounds. Anderson (inf.) says that in waters south of Hawaii, in such places as Washington Island (central Pacific), where a shallow lake fills the centre of the island, the inhabitants stock the lake or lagoon with various young fish from the reef waters, among them *ulua*. After *ulua* have spent several weeks in the sandy-bottomed lake, the black and white *ulua* are indistinguishable.

However, Hawaiians are familiar with them as dark and light-coloured *ulua*. Describing them from casts in Bishop Museum, *ulua lauli*: *C. lugubris* (cast 378) is about 28 inches long. A report from Molokai is that they average 60 pounds, the largest taken there being 97 pounds in weight. In the cast, the colour is very dark grey green all over, slightly lighter below, the eye surrounded by a pale green area. All *ulua* are fighters when on the line, but this is the greatest fighter of them all. They frequent certain currents and are seldom caught elsewhere.

Ulua aukea (or *kea*): *C. ignobilis*, is represented in two casts in Bishop Museum. Cast 364 is about 3½ feet long, the colour yellowish green, considerably mottled, darkest on the highest part of dorsal region; upper edge of pectoral fin touched with dark tones, possibly brownish; tail fin likewise darkest on anterior edges, other fins dark in tone. The other cast (381) is about 28 inches long, colour is a mottled light green or blue, whitish below—the effect pastel. Fins are whitish, touched at edges with blue. Watson (inf.) says, " In Kane'ohe Bay it stays inside the reef, reaches 200 pounds in weight, and is a common fish in the market."

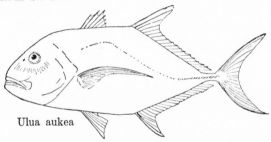

Ulua aukea

Ulua 'ele'ele (black). This may be *C. elacate* (Jordan and Evermann), which is even darker than *C. lugubris*. It is more streamlined than *C. lugubris*, and is less commonly seen. Cast 15, in Bishop Museum, shows a fish about 30 inches long, colour of head brown, almost black, the rest of the body almost as dark in tone, with an under colour of green, lighter colouration below the lateral line, fins all very dark; eye yellow, an eye shield extending posterior of the eye as far again as the diameter of the eye. Of *ulua 'ele'ele*, Watson says " pure black, except for white spots on head, usually about 3 feet long, but sometimes twice as long."

The smaller *ulua* include:—

Ulua paopao (ridged, or rippled), sometimes called *ulua kani'o* (striped): *Gnathanodon speciosus* Forskal. Length is about 20

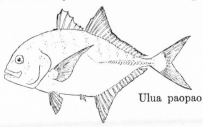

Ulua paopao

inches, weight 18-20 pounds (Galbraith, inf.). Jordan and Evermann
(40) furnish a coloured plate (No. 12). Cast 9 in Bishop Museum is
more vivid in colouration, greenish above lateral line, yellow below,
with the distinctive marking of this *ulua*—eight or nine green vertical
bands. Ventral and anal fins are yellow, dorsal green with yellow at
the outer edge, tail fin yellow, touched and edged with brown. This
ulua is recognized among Hawaiians as the best for eating raw. It
may be cooked, however. One habit characterizes the *paopao*. As it
dives for food it leaves its tail waggling above the surface of the
water. (Joseph Kawelo, inf.)

Ulua nukumoni (pearly snout): *Caranx melampygus* Cuvier.
Cast 14 in Bishop Museum is about 18 inches long, but informants say

Ulua nukumomi

it reaches a little greater length; colour of cast is grey green, mottled;
fins all very dark brown except pectoral, which is grey green, with
brownish inner edge; gill shield has touches of brown.

Ulua kihikihi (angled, cornered), also called *mahai* (tapered),
hulipū (turned over completely, as a canoe that is bottom up):
Blepharis ciliaris (Bloch). Breder (9, p. 137) says, "Formerly
thought to be an independent species, but now believed to be the

Ulua kihikihi

young of a rather different looking fish of large size . . ." Hosaka
(37, p. 143, fig. 112) offers the belief that this fish in the adult stage
is the one popularly known as the "silver *ulua*," and says it is "bright
silver with a bluish tint." Kondo says some Japanese fishermen call
it the mirror fish, because of its dazzling whiteness. Jordan and
Evermann (40, p. 200), says, "The changes due to age are surprisingly
great . . ." There is a cast (No. 4) in Bishop Museum of this fish at
its young stage. No transverse bands are evident but otherwise it

agrees with the description by Jordan and Evermann (40, p. 200-201),
" Body oval, much compressed . . . first rays of dorsal and anal
filamentous, exceedingly long, in the young much longer than body,
becoming somewhat dusky above and showing very distinct when
held at certain angles . . . black spot on base of highest portion of
dorsal . . ." In the cast there is a dark spot on the lowest part of the
anal fin also; colour of the whole fish in the cast is silvery, fins pale
yellow with dark thread fins, touches of dark blue along dorsal spines,
outer edges of ventral fins, and base of pectoral; eye large, yellow
with dark pupil and outer ring. Body is five inches long, length of
anal threads 6½ inches.

There seems to be lack of agreement between fishermen as to the
ulua kihikihi. Some claim the finlets drop off as the fish matures;
some say they do not. Perhaps the *Blepharis* loses the long finlets.
Another fish, *Scyris indica* Rüppell (Hawaiian name not found), is
quite similar and a cast (3392) in Bishop Museum shows the fins have
retained their filamentous character at what must be adult stage. The
cast is very light in colour, faintly rose, merging into violet at the
dorsal region; pectoral fins pale greenish, tail fin greenish, slightly
brown; finlets dark; a dark smudge over the eye, and another at the
upper edge of the gill shield; ventral fins dark. This fish has been
reported and described many times—at length by Fowler (26, pp. 151-
52).

Other Hawaiian names collected are the *ulua omilu,* doubtless an
omilu, the *ulua kukaenalo,* doubtless the freckled *omilu,* and *ulua
kahauli* (dark stripe), for which no guess is offered. Two informants,
one from Ka'u in Hawaii, the other from Molokai, also speak of the
ulupō. One says it is commonly found at Huiha, Kona, Hawaii, and
resembles the *papiopio ulua,* that is it is small in size. The other
informant says it has a prominent nose, and vertical stripes, and grows
to about 20 pounds.

The *ulua* group of fishes is a particularly difficult one to classify.
One ichthyologist, J. T. Nichols (pers. comm., letter of 11/5/48),
says:—

> " The principal Hawaiian food fishes in this family are
> probably four species of the genus *Caranx.* These Weber and
> de Beaufort (*Fishes of the East Indian Archipelago*) and I call
> *Caranx ignobilis, sexfasciatus, melampygus,* and *stellatus.*
> *C. melampygus* averages about a foot long, and probably does
> not reach the large size of the others (which average two feet
> in length). It is the most abundant in shore waters near Hono-
> lulu. It is a white fish with black fin tips. *C. ignobilis,* and
> *sexfasciatus* (normally, except when large) are also white fishes
> and, though technically different, the two most likely to be con-
> fused by fishermen. *C. stellatus* is a darker fish.

> " Jordan and Evermann describe three of these four fishes
> under different names; their *ignobilis* is the same. They use
> three names for *sexfasciatus,* namely *rhabdotus* (young),
> *marginatus* (moderate) and *elacate* (large). They use
> *melampygus* for *stellatus.*

> " Fowler synonymizes *melampygus* and *stellatus* under

ascensionis, but I do not agree. The fishes he figures as *ascensionis* and *lugubris* I am familiar with as two similar black fishes. The former is found in a special habitat, about isolated rocky islets, and reported to be unwholesome—I do not know how generally so. The names *lugubris* and *ascensionis* likely rightly belong to the same fish. However, having seen only photographs of Hawaiian "*lugubris*" I can not place it with certainty, and it seems as or more likely that it is large *ignobilis,* possibly those from a particular habitat.

"*Ulua lauli* suggests *stellatus* rather than *lugubris.* C. *ignobilis sexfasciatus* (normal colour) and *melampygus* may all be *ulua aukea,* but it would be surprising if there were not a special Hawaiian name for both *melampygus* and *stellatus.* Perhaps Jordan and Evermann are correct that *stellatus* (their *melampygus*) is *omilu.*"

Ulua were eaten raw or cooked. The preference for *ulua paopao* for eating raw has already been mentioned. If large, *ulua* were usually baked, if small broiled. The eyes were well liked, stuffed into the belly before the fish was placed in the *imu.* The savoury liquid around the eyeball was the delicacy, the eyeball itself becoming hard and unpalatable when cooked.

Umaumalei (*umauma:* chest; *lei:* garland for the neck). Similar to but darker than the *palani* or *pualu;* has bright orange-red spots around the gills and side fins (pectorals), and at the base of the caudal fin where the spike is set. (Mrs. Pukui.)

Ume (to attract, draw), another name for *kala.*

Uouōa (*wowōa,* or *uōa*): *Neomyxus chaptalii* (Eydoux and Souleyet), and false mullet, or false *'ama'ama* (in Hawaii). Mrs. Pukui points out the chief difference between the two fish, " In the *uouōa,* the mouth is not as wide and is slightly more tapering, the pectoral fins are creamy in colour instead of silvery." The young are called *pua uouōa.*

Uouōa

Description: Length, up to 9 inches; colour in alcohol, " dull olivaceous above with silvery reflections gradually fading into lighter and becoming white on belly; top of head and snout brown . . . dorsal, caudal and pectorals dusky, pectorals the darkest; other fins pale." (40, pp. 140-141). Scales large. Mainland (62, pp. 68-69) has described the *uouōa:*—

Colour, back is dull, silvery olive-grey, colour not evenly diffuse, but in small specks; from the dorsal portion of the side to the belly, the colour gradually fades to a silvery white. Small yellow spot on the upper base of the pectoral fin; cheeks olive-grey, snout and lips before eyes olive-grey. Dorsals and caudal dusky, the latter with a dark diffuse border along the margin of the fork; ventrals and anal white, upper portion of pectorals dark, lower portion white. Size, may reach 12 inches or more; habitat: the normal habitat . . . seems to be along the shore . . . but it may enter estuaries and ascend the lower portion of the streams.

Eaten raw, or cooked in *ti* leaves; taste is the same as the mullet.

This is one of the fish the head of which sometimes contained a substance that caused nightmare or wakefulness to the consumer. The head was often thrown away, rather than take this chance. (Wiggin, inf.) This quality is mentioned in the tale of Kamiki (75, 23, Feb. 15, 1911) :—

> On this beach (where Kamiki killed the ghosts) the mullet, *uoa* and *weke* have bitter heads and gills, like the taste of tobacco leaves . . . Therefore, if you should eat a mullet, *uoa*, or *weke* fish, thrown away a bone from the fish you are eating, and say, "Here is your share, O Pahulu." (Pahulu was the chief of the ghosts, and they sometimes dwelt in the heads of these fishes, making them bitter.)

This same character, Kamiki, chanted at one point of the tale of various fish:—

> The travelling *'ama'ama* of Kapuualii,
> The *uoa* of the sandy beach of Kuula,
> The *weke-lao* of Na-maka-o-Kane
> Now lie in a long container,
> In the gourd calabash of my grandmother . . .
> (75.23, Feb. 15, 1911)

Upapalu (soft), sometimes *upapalu-maka-nui* (big-eyed) : *Apogon frenata* (Valenciennes) ; *A. maculifera* (Garrett). Watson thinks *upapalu-maka-nui* is the name given to the larger of the two species, probably *A. maculifera*.

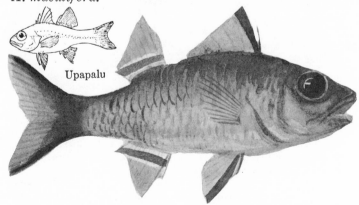

Upapalu

Description: Nakuina says the *upapalu* is "hand length when mature; colour is pinkish silver." J. and E. (40, pp. 214-215) says of the *A. frenata* (sp. 165, JE), that it is "pale red, two longitudinal pearly lines on body . . . ," and the length as 3 to 5.5 inches, "reaches a length of 6 inches." Watson says 3 inches is the common size. *A. maculifera* is described as about 6 inches; colour, pale purplish grey, belly pale orange, head dusky reddish orange, with purplish tinges; six or seven rows of dark spots longitudinally on body, fins of many colours. (40, pp. 212-213, sp. 163.) Scales are large. There are other Hawaiian species of *Apogon*, and perhaps all of them are *upapalu*. They differ greatly in markings; ground colour usually reddish silver, or red.

The meat is sweet, soft and tender. It is good raw, broiled, or wrapped in *ti* leaves, then broiled. Flesh is white, not many bones. (Nakuina, inf.)

It is caught on moonlight nights, the first fish to bite. They come to the surface in great numbers; common in Puna and Kaʻu. (Linolau Leong, inf.)

Ūʻū: *Myripristis sealei* (Jenkins) ; *M. murdjan* (Forskal) ; *M. chryseres* (Jordan and Evermann) ; *M. argyromus* (Jordan and Evermann) ; *M. multiradiatus* (Günther) ; *M. praslinus* (Cuvier) ; squirrel fishes.

Ūʻū

Description: Length varies from 5 to 9 inches, occasionally longer. In Kaneʻohe Bay, they are usually 3 to 4 inches long (Watson). The predominating colour for all of them is red, some of more intense colour than others, sometimes with silvery and whitish areas, sometimes the colour is reddish brown or bronzy; fins also red or rosy. All have big eyes and large scales. (40, 149-155.)

This is a delicious fish, good to eat raw or broiled. Being very difficult to skin, it is easiest to broil and eat the flesh from the skin. For eating raw the fish is cleaned of internal organs, and the dorsal and anal fins are pulled away, ripped out. (Linolau Leong.)

The large eyes of the squirrel fishes are luminous at night, giving the effect of a group of stars. They are caught at night, by nets, or by hook and line; they bite vigorously. In the daytime they hide in caves in the coral. They are found both in schools and singly. The Japanese in Hawaii are fond of them; for them they take the place of carp. It is an expensive fish in the market, not being plentiful along many reefs. (Watson.)

Uwiwi, see *uiui,* under *humuhumu.*

Uawau, listed (14, p. 441), may refer to *owau,* an *'o'opu.*

Walu (*wolu* in Ka'u and Puna, *walu* in Kane'ohe, Oahu):

Ruvettus pretiosus (Cocco) ; the oilfish. (Called palu in most of Polynesia).

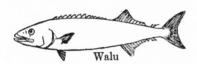

Walu

Description:--

" The average size is about 3 to 4 feet, and weight 40 to 60 lbs. . . . A full-grown 'Palu' would weigh up to 150 lbs., and be 6 ft. long; it being by no means a thick fish . . . In place of scales it possesses a tough black skin, thickly covered with bright silvery and small horny excrescences growing in the same manner as the feathers of a French fowl—that is, these scales or whatever you can call them, curl upwards, and feel loose to the touch. The most peculiar feature of the 'Palu' are the enormous eyes; the jaws are toothless; the fins resemble those of a Jew fish.

" . . . every portion of it is edible; the head, bones, and fins, when cooked turning into a rich mass of jelly. The flesh of the 'Palu,' if left uncooked, never putrefies; it simply dissolves into a colourless and odourless oil . . . Its almost immediate effect on the bowels . . . not to be too coarse . . . the fish that makes you obey the call of nature in double quick time. (Ellice Island expression.) . . . Caught in 80 to 120 fathoms." (3, pp. 200-201.)

When this description was read to Mrs. Pukui, she said every part of it fitted the Hawaiian *wolu* exactly. In Hawaii it was eaten fresh or dried, or broiled after drying; not eaten raw. Wiggin says: " Flesh has much oil, bones are soft and can be eaten along with the flesh. It is too rich in oil to eat of heartily. A little eaten at a meal is enough; used as a cathartic it is more effective than castor oil, but without the distressing cramping effects. When dried, the flesh becomes yellow."

It was caught off Kona, Hawaii, a favourite place for deep sea fishing. It was sold in the Honolulu market, before World War II, always dried, and brought 80 cents a pound. As elsewhere in Polynesia, it is a prized fish. (Mrs. Pukui.)

Weke (to open), certain species of the *Mullidae*, surmullets.
Hawaiian names collected are *weke ‘ula* (red), *moelua*
(translatable in more than one way, no decision can be
made), *nono* (bright red), *la‘ō* (sugar cane leaf), *pueo*
(owl), *pahulu* (chief of the ghosts), or *ahulu* (nightmare), *a* or *a‘a* (staring), *koā‘e* (the tropic bird),
ke‘oke‘o (white). Of these the most commonly used are
weke ‘ula, pueo, a (or *a‘a*) and *pahulu*. The young of
all are called *oama* (finger length to about 6 inches).

Weke ‘ula: Mulloidichthys auriflamma (Forskal).

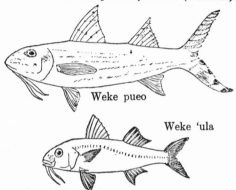

Weke pueo

Weke ‘ula

Description: Average length about 9 inches (Watson), sometimes reaches 14 inches (40, p. 250); colour, "upper half of head,
nape, and back rosy red, richest on head, lower half of head white
with very light rosy wash; side with pale yellow band a scale wide ...
lower two-thirds of side white with a light rosy wash, fins all pale
rosy, except pectoral which has a slightly lemon-yellowish wash."
(40, p. 250.) Kawelo says the colour effect of a school of *weke ‘ula* is
not red but yellow. They show themselves red when taken out of the
water. (Or is it possible that the colour changes to vivid red when
they are drawn from the water?)

Weke ‘ula are taken inshore usually, and are the best eating of
the *weke*, the favourite preparation being broiling in *ti* leaf wrapping.

Weke pueo (owl): *Upeneus arge* (Jordan and Evermann). There is a
striking similarity between the stripes on the tail of this fish and the bars
of the Hawaiian owl's feathers.

Description: Length, 8-12.5 inches (cast 467 in Bishop Museum
measures a little over 14 inches); colour "pale green, changing to
white below, edges of scales on back and down lateral line purplish
brown, giving the appearance of three rather distinct stripes of
purplish brown, with greenish centres on the scales; side with two
broad yellow stripes ... cheek white with some rosy, lower jaw
white, barbels yellow, dorsal fins pale, each crossed by two or three
brownish rosy bars, caudal white, upper lobe with broad brownish red
bars ... lower lobe ... similar, but ... broader black bars ... (other
fins) pale ...; shallow waters, quiet shores." (40, 264-65.) Watson
gives a different picture entirely, saying that the entire body is
covered with little black spots, most prominent on the tail, and adds
it is most often found on sandy flats, running in schools. Kawelo

finds it in deep water most frequently. Cast 467 in Bishop Museum agrees in colour with J. and E. fairly well though the colour stripes along the body are more rosy than purplish brown, and one in the middle is clear yellow; spines of the second dorsal are touched with the same colours as the "owl" bars of the tail fin, as if in faint imitation; barbels are yellow.

weke a (*a'ā*) (staring) (synonym is *ke'oke'o*, according to Alona): *Mulloidichthys samoensis* (Günther).

Description: Length, up to 13 inches; colour, back greenish olive, a yellow band one scale wide from the eye to tail, below this band the colour is "white with two very faint yellow lines, belly white; fins all whitish, the spinous dorsal yellow on anterior part; soft dorsal and caudal washed with yellow." (40: 253.) Kawelo says "whitish with yellow stripes lengthwise."

Weke pahulu (chief of the ghosts), *Upeneus arge* (Jordan and Evermann), also called *ahulu* (nightmare). The flesh of the head of some *weke* has a poisonous quality. Those who eat it have restless sleep or nightmare in which the sensation is one of having lost balance, and especially one of feeling that the head is lower than the feet and it is impossible to get it back to level as one is lying down. A report by John M. Wilson on the *weke pahulu* says that (42: 674):—

Eating the head of this fish produces a sort of delirium . . . At one time in Molokai 30 or 40 Japanese labourers . . . ate the heads (and bodies) of many examples and were mentally paralyzed at the time. Mrs. Wilson once attended a function at which this fish was served. All members of the party had weird visions, some of them wandering about the house all night . . . It is agreed that the poison lies in the brain. A fresh example was turned over to Dr. Nils Larsen, director of the Queen's Hospital. He fed the brain to a cat, which at once went crazy, but recovered, as in fact, all cases soon recovered. Dr. Larsen fed other species and the flesh of *U. arge* to cats but with no results. In the winter he fed the brain of this species to cats and they were not affected.

The question naturally comes to mind, "Is the poison derived from some food of the *weke pahulu* which is present only at certain seasons?" And also, "Does this *weke* eat seaweeds (?) which other *weke* do not eat, or has the *weke pahulu* the special ability to segregate this substance in its brain case; and of what value is it to the fish?"

Weke caught off the island of Lanai and Molokai are most apt to be poisonous; Alona says *pahulu* is the Lanai name for *weke ā* or *ke'oke'o*; Kawelo says it is the Lanai name for *weke pueo*; Watson says that in Kane'ohe Bay there are three *weke*, the *koā'e* ("white with small yellowish spots, fins light yellowish, about 6-15 inches long"—which may be the *weke ā* to other fishermen"), the *'ula* and the *pahulu*, which he describes as distinguished by "yellow spots on body, black spots on tail." This must be the *weke pueo*, the exact use of such terms as spots, splotches, lines, which are sometimes true lines and sometimes a series of dots in line, being difficult.

Weke nono (bright red). This may be the name of a red *weke*, a cast of which is in Bishop Museum (No. 468), labelled *Upeneus chrysonemus*, J. and E., which Fowler discusses (26, p. 229-230) and identifies as *Upeneus taeniatus* Kner.

Description: Cast a little over 7 inches long; colour, back deep carmine red, continuously so along middle of back; just posterior to head and at base of soft dorsal the red colour is most intense; rest of body yellow, with flecks of red at edges of scales; head is red above yellow below, with blue streaks from mouth to eye and blue dots on the gill shield. Lines of blue dots found also on the caudal peduncle. Fins vary in colouration, first dorsal and ventral yellow with bright red spines; second dorsal yellow above, with lavender stripes, purplish below; anal yellow with lavender stripes, tail fin mostly red but tinged with yellow at outer edge, barbels yellowish.

Weke lā'ō. No Hawaiian has described this *weke.* The meaning (sugar-cane leaf) may indicate that it is pale greenish, hence another name for *weke ā.*

All *weke* have large scales. They are usually found inside the reefs, sometimes in the deep waters outside but near the reefs.

Weke are popular fish as food. The *oama* (young) are delicious eaten raw after being salted a few minutes, or dried. To remove the scales, the *oama* were put into a large container with pebbles and sand, stirred until the scales were loosened or rubbed off, then rinsed in sea-water. Large *weke* scaled by scraping. Full-grown *weke* are sometimes eaten raw, but usually cooked, broiled in *ti* leaves over hot coals.

Both red and light-coloured *weke* were popular as offerings to the gods, chosen according to the demands of customs, red for certain occasions, white for others. The meaning of the term *weke* (to open) also gave value to this fish in sorcery, as a priest might offer it with a prayer to open or release something, such as evil thoughts, preparatory to forgiveness, or a prayer to open the door of mystery so as to reveal the truth (Mrs. Pukui).

The harmful substance in the head has a legendary explanation in the story of Pahulu, chief of the ghosts. One version is as follows (75.41) :—

> Kaululaau sat in the *milo* tree with a flat stone. He saw Pahulu peer into the spring, for the light of the moon shone fully upon it. As Pahulu stooped to dip up some water, Kaululaau pushed aside some of the *milo* leaves. The moon cast his reflection in the water. He made grimaces, and when Pahulu saw the reflection making faces (he was deceived), and he dived to catch him. As soon as Kaululaau saw that Pahulu's whole body had gone into the water he threw the stone down. The spark of life went out of Pahulu's body and he died. His spirit leaped into the sea, and that is why the people who eat *weke* are troubled with nightmares.

A briefer mention of this legendary incident is worth adding:—

" Then Kaululaau went and sat over the pool where he dropped a stone on Pahulu, killing him. Pahulu died but he lived on in the itching caused by certain fish. That is why there is itching in the head of the *weke* and the itch is always there in the fish caught off Lanai to this day." (75.4.)

> Still another version is in the tale of Kamiki, a legendary hero.
>
> Kamiki ran swiftly to the beach of Makalawena, by the sand dunes called Kapuualii and Muula . . . On this beach the mullet, *uoa* and *weke* have bitter heads and gills, like the taste

of tobacco leaves. It is said that this was because Kamiki caught the ghosts here. Therefore if you should eat *'ama'ama, uoa,* or *weke* fish throw away a bone from the fish . . . and say, " Here is your share, Pahulu." Then when you eat it there will be no bitterness from the head to the tail. (75.23, Ke Au Hou, 2/15/1911.)

The recorder of this story is puzzled, for he adds, " This is strange and mystifying, but your writer knows this to be so. If you should ask the natives of the place they will verify this."

Wela, see *puhi wela.*

Welea: Trachinocephalus myops (Schneider), lizard fish.

Welea

Description: Cast 122 in Bishop Museum about 9 inches long. Head and dorsal region lavender, with irregular brownish outlines forming circles and mottled areas; body banded lengthwise in lavender, with brownish edges over a base of pale yellow; fins yellowish with touches of lavender. Body plump; there is a little depression or valley between the eyes. The markings vary from those described by Jordan and Evermann (40, p. 62), " pale greyish . . . side of back with crossbars of dull yellow, each edged with darker olive . . . below these are two faint streaks."

Thompson notes on the label of the cast that this is a reef fish.

Welehu: a deep sea fish resembling the *hauliuli.*

Weuweu. Listed (59 a).

Wolu, see *walu.*

BIBLIOGRAPHY.

1. ANDERSON, William. *An account of some poisonous fish in the South Seas . . . on board . . . " Resolution " . . .* R. Soc. London, Trans., 66:554-574, 1776.

2. ANDREWS, Lorrin & PARKER, H. H. *A dictionary of the Hawaiian language.* Honolulu, 1922.

3. BECKE, Louis, in WAITE, E. R. *The mammals, reptiles, and fishes of Funafuti.* Australian Mus. Mem., 3:165-201, 1897.

4. BECKLEY, E. M. *Hawaiian' fisheries and methods of fishing with an account of the fishing implements used by the natives of the Hawaiian islands.* Honolulu, 1883. (Reprinted in U.S. Fish Com., Bull. 6:245-256, 1886.)

5. BECKWITH, M. W. *Hawaiian mythology.* New Haven, 1940.

6. BENNETT, F. D. *Narrative of a whaling voyage round the globe . . . 1833 to 1836 . . .* Vols. 1-2, London, 1840.

7. BENNETT, W. C. *Archaeology of Kauai.* Bernice P. Bishop Mus., Bull. 80, 1931.

8. BLEEKER, Pieter. *Atlas ichthyologique des Indes Orientales Neerlandaises.* Tomes 1-9, Amsterdam, 1862-1878.

9. BREDER, C. M. *Field book of marine fishes of the Atlantic coast.* New York, 1929.

10. BRIGHAM, W. T. *Director's report for 1901.* Bernice P. Bishop Museum, Occ. Papers, Vol. 1 (5) : 1-31, 1902.

11. BRYAN, W. A. *Natural history of Hawaii.* Honolulu, 1915.

12. BUCK, P. H. *Hawaiian shark-tooth implements.* Bernice P. Bishop Mus., Bull. 180: 27-41, 1943.

13. COBB, J. N. *The commercial fisheries of the Hawaiian islands.* U.S. Fish Comm., Rept. 1900/01: 383-499.

14. COBB, J. N. *The commercial fisheries of the Hawaiian islands in 1903.* U.S. Fish Comm., Rept. 1903/04: 433-512.

15. COOK, James. *A voyage to the Pacific Ocean . . . 1776-1780.* Vols. 1-3, London, 1784.

16. CRAWFORD, David L. *Hawaii's crop parade.* Honolulu, 1937.

17. DAKIN, W. J. *Migration and productivity in the sea . . .* Australian Zoologist, 7:15-31, 1931.

18. DUHAUT-CILLY, A. *Voyage autour du monde . . . 1826, 1827, 1828, 1829.* Vols. 1-2. Paris, 1834-35.

19. DUMERIL, Auguste. *On venomous fish.* Ann. Mag. Nat. Hist., Ser. 2, Vol. 20: 153-167, 1867.

20. EDMONDSON, C. H. *Reef and shore fauna of Hawaii.* Bernice P. Bishop Mus., Special Publication 22, 1933.

21. ELLIS, William. *An authentic narrative of a voyage performed by Captain Cook and Captain Clerke . . . 1776-1780.* Vols. 1-2. London, 1782.

22. EMERSON, J. S. *The lesser Hawaiian gods.* Haw. Hist. Soc., Paper 2, 1892.

22a EMERSON, N. B. *Unwritten literature of Hawaii.* (Bur. Am. Ethn., Bull. 38, 1909.)

22b EMORY, K. P. *The island of Lanai, a survey of native culture.* (Bernice P. Bishop Museum, Bull. 12, 1924.)

23. FARRINGTON, S. K. *Pacific game fishing.* New York, 1942.

24. FIELD, H. G. *Game fishing in Hawaiian waters.* Haw. Ann. 1917: 87-93.

25. FORNANDER, Abraham. *Fornander collection of Hawaiian antiquities and folklore.* Ser. I-III: Bernice P. Bishop Mus., Mem. Vol. 4-6, 1916-20.

26. FOWLER, H. W. *Fishes of Oceania.* Bernice P. Bishop Mus., Mem. Vol. 10; 11: Nos. 5, 6, 1928, 1931, 1935.

27. FREYCINET, L. C. D. de. *Voyage autour du monde . . . l'Uranie . . . 1817-1820 . . ., Zoologie, Atlas by Quoy et Gaimard.* Paris, 1824.

28. GREEN, L. S. & BECKWITH, M. W. *Hawaiian household customs.* Am. Anthropologist, N.S., Vol. 30: 1-17, 1928.

29. GREY, Marion. *Regalicus, Lampris, Mola and Ranzania . . .* Chicago Nat. Hist. Mus., Bull. 16 (11-12) : p. 5, 1945.

30. GRIMBLE, Sir Arthur. *Migrations of a Pandanus people.* Poly. Soc., Mem. Vol. 12, 1933.

31. GUDGER, E. W. *Poisonous fishes and fish poisonings . . .* Am. Jour. Trop. Med., 10 (1) : 43-55, 1930.

32. GUNTHER, A. C. L. G. *Andrew Garrett's Fische der Südsee,* Band 1. Jour. Mus. Godeffroy, 2, 1873-75.

33. HANDY, E. S. C. *Polynesian religion.* Bernice P. Bishop Mus., Bull. 34, 1927.

34. HANDY, E. S. C. & PUKUI, M. K. "*Ohana,*" the dispersed community of "*Kanaka.*" Manuscript in Bernice P. Bishop Museum.

35. HANDY, E. S. C., PUKUI, M. K. & LIVERMORE, Katherine. *Hawaiian physical therapeutics.* Bernice P. Bishop Museum, Bull. 126, 1934.

36. HENSHAW, H. W. *Birds of the Hawaiian islands.* Honolulu, 1902.

37. HOSAKA, E. Y. *Sport fishing in Hawaii.* Honolulu, 1944.

38. JORDAN, D. S. (Letter to Captain U. Sebree; informal report on Samoan fishes, poisonous and edible, dated Aug. 2, 1902, Tutuila.)

39. JORDAN, D. S. *A guide to the study of fishes.* Vols. 1-2, New York, 1905.

40. JORDAN, D. S. & EVERMANN, B. W. *The aquatic resources of the Hawaiian islands.* U.S. Fish Comm. Bull. Vol. 23, Part 1, 1903.

41. JORDAN, D. S. & EVERMANN, B. W. *Preliminary report on an investigation of the fishes and fisheries of the Hawaiian islands.* U.S. Fish Comm. Rept., 1900-1901: 353-382.

42. JORDAN, D. S., EVERMANN, B. W. & TANAKA, Shigeko. *Notes on new or rare fishes from Hawaii.* Calif. Acad. Sci., Proc., Ser. 4, Vol. 16: 649-680, 1927.

43. JORDAN, D. S. & JORDAN, E. K. *A list of the fishes of Hawaii . . .* Carnegie Mus., Mem. Vol. 10: 1-92, 1922-1925.

44. JORDAN, D. S. & SNYDER, J. O. *Description of nine new series of fishes contained in museums of Japan.* Imp. Univ. of Tokyo, College of Sci., Journal, Vol. 15:301-311, pl. 15-17, 1901.

45. *Journal of the missionaries (to the Sandwich Islands), April 2, 1820.* Missionary Herald, Vol. 17:115-116, 1821.

46. JUDD, H. P. *Hawaiian proverbs and riddles.* Bernice P. Bishop Mus., Bull. 77, 1930.

47. KAMAKAU, S. M. *Moolelo o Hawaii.* Manuscript in Bernice P. Bishop Museum.

48. KAMAKAU, S. M. *Moolelo o Kamehameha.* Manuscript in Bernice P. Bishop Museum.

49. KELIIPIO, L. D. *Hawaiian fish stories and superstitions.* Haw. Ann. 1901: 110-114.

50. KELSEY, Theodore. *Notes collected.* Manuscript in Bernice P. Bishop Museum.

51. KELLY, H. L. *Problems connected with the fishing industry in Hawaii.* Haw. For. and Agriculturist, 27 (1): 9, 1930.

52. KEPELINO (Keauokalani Kahoalii, known as Kepelino). *Ka moolelo o na iʻa Hawaii.* Manuscript in Bernice P. Bishop Museum.

52a KEPELINO. *Kepelino's traditions of Hawaii, edited by M. W. Beckwith.* Bernice P. Bishop Museum, Bull. 95, 1932.

53. KRUMHOLZ, L. A. *Northward acclimatization of the western mosquito fish.* Copeia, 1944: 82-86.

54. LA MONTE, F. R. *North American game fishes.* Garden City, 1945.

55. LA MONTE, F. R. & MARCY, D. E. *Swordfish, sailfish, marlin and spearfish.* Ichthyological Contributions of the Int. Game Fish Assn., 1 (2): 1-24, 1941.

56. LARSEN, N. P. *Tetrodon poisoning in Hawaii.* Pacific Sci. Congress, 6th Proc., Vol. 4: 417-421, 1942.

57. LEE, R. K. C. & PANG, H. Q. *Ichthyotoxism—fish poisoning.* Am. Jour. Trop. Med., 25 (3): 281-285, 1945.

58. LESSON, R. P. *Voyage autour du monde . . . 1822-25.* Vol. 1-2. Paris, 1839.

59. Lists of Hawaiian fishes: Manuscript lists in Bernice P. Bishop Museum: a—Kalakaua; b—Liliuokalani; c—C. M. Hyde; d—Edgar Henriques; e—E. H. Bryan, Jr.; Newspaper lists: f—Ke Au Hou, 6/15/1910; g—Ka Hae Hawaii, 4/4/1860; h—Ka Lahui Hawaii, Aug. 10 and 26, 1899; i—list from source unknown.

59a McALLISTER, Gilbert. *Archaeology of Kahoolawe.* Bernice P. Bishop Mus., Bull. 115, 1933.

60. McALLISTER, Gilbert. *Archaeology of Oahu.* Bernice P. Bishop Mus., Bull. 104, 1933.

61. MADISON, N, H. *Florida fishes.* Cleveland Mus. of Nat. Hist., Pocket Nat. Hist., No. 5, 1936.

62. MAINLAND, G. B. *Goboidea and fresh water fish on the island of Oahu.* Manuscript in University of Hawaii Library.

63. MALO, David. *Hawaiian antiquities.* Bernice P. Bishop Mus., Sp. Pub. 2, 1903.

64. MANBY, Thomas. *Journal of Vancouver's voyage to the Pacific Ocean, 1791-1793.* Manuscript copy of the Hawaiian entry, in Bernice P. Bishop Museum.

65. MANU, Moke. *Kuula, the fish god of Hawaii; translated by M. K. Nakuina.* Haw. Ann., 1901: 114-124.
66. MASSEE, E. K. *Fishing rights in Hawaii.* Honolulu, 1926.
67. MAUNUPAU, T. K. *Notes on off-shore fishing in Hawaii.* Manuscript in Bernice P. Bishop Museum.
67a MAUNUPAU, T. K. *A visit to Kaupo, Maui.* (Ka Nupepa Kuokoa, July 27, 1922, translated by M. K. Pukui.)
68. MAXWELL, C. N. *Malayan fishes.* Royal Asiatic Soc., Straits Branch, Journal, No. 84, 1921.
68a MEARES, John. *Voyages made in the years 1788 and 1789 . . .* London, 1790.
68b MENZIES, Archibald. *Hawaii Nei 128 years ago.* Honolulu, 1920.
69. NICHOLS, J. T. & LA MONTE, F. R. *Differences in marlins.* Ichthyological Contributions of the Int. Game, Fish Association, 1:1-8, 1941.
70. NICHOLS, J. T. & LA MONTE, F. R. *How many marlins are there?* Natural History, 36: 327-330, 1935.
71. PAULAY MARIN, Don Francisco de. *The Marin journal (edited by Maude Jones), entry of July 31, 1814.* Paradise of the Pacific, 49 (9): 2, 1937.
72. PIETSCHMANN, Victor. *Hawaiian shore fishes.* Bernice P. Bishop Mus., Bull. 156, 1938.
73. POGUE, J. F. (Rev. J. F. Pokuea). *Ka moolelo Hawaii.* Honolulu, 1858. Manuscript translation into English of the pre-European section, by M. K. Pukui.
74. PUKUI, M. K. *Hawaiian beliefs and customs during birth, infancy, and childhood.* Bernice P. Bishop Mus., Occ. Pap., Vol. 16 (17): 357-381, 1942.
75. PUKUI, M. K. Translated excerpts from Hawaiian newspapers:—
    1. (No title). Ke Aloha Aina, 9/28/1895.
    2. *Ancient beliefs.* Ka Nupepa Kuokoa, 12/8/1865.
    3. *Answers to questions.* Ka Hoku o Hawaii, 5/22/1940.
    4. *Eleio.* Ke Au Hou, 10/31/1863.
    5. *Haina nane.* Ka Nupepa Kuokoa, 4/23/1925.
    6. *Haina nane.* Ka Hoku o Hawaii, 5/22/1940.
    7. *History of Hawaii nei.* Ka Nupepa Kuokoa, 9/23/1865.
    8. *Hookumu ana o na pae moku.* Ke Au Hou, 2/8/1911.
    9. *Huakai makaikai i'a Hawaii akau.* Ka Lahui Hawaii, 10/18/1877.
    10. *I'a hou ma Niihau.* Ka Hae Hawaii, 2/17/1858.
    11. *Ka hana kuhikuhi no ka lawai'a ana.* Ka Nupepa Kuokoa, 5/23/1902.
    12. *Ka i'a alalauwā maloko o ka awa o Kou.* Ke Aloha Aina, 8/22/1919.
    13. *Ka i'a pae.* Ka Nupepa Kuokoa, 9/4/1870.
    14. *Ka'ao o Puakaohelo.* Ke Au Okoa, 1/5/1871.
    15. *Ka nane hūna a Kauikeaouli.* Ke Aloha Aina, 8/31/1895.
    16. *Ka moolelo o ka oihana lawai'a.* Ka Nupepa Kuokoa, 5/23/1912.
    17. *Kaehuiki-mano-o-Pu'uloa.* Ke Au Hou, 4/26/1911.

18. *Kaililauokekoa.* Ka Hoku o Hawaii, 10/1/1908.
19. *Kamaakamahiai.* Ke Aloha Aina, 6/3/1911.
20. Kamakau, S. M. *Story of Kamehameha.* Ke Au Okoa, 11/17/1870.
21. Kamakau, S. M. *Story of Hawaii.* Ke Au Okoa, 1/6/1870; Ka Nupepa Kuokoa, 12/23/1869.
22. *Ke Alalauwā.* Ka Nuhou, 8/12/1873.
23. *Tale of Kamiki.* Ke Au Hou, 2/15/1911; 3/1/1911; Hoku o Hawaii, 3/19/1914; Hawaii Holomua, 11/13/1912.
24. *Keamalu.* Hoku o Hawaii, 1/3/1917.
25. *Kua iali'i.* Kuokoa Home Rula, 6/17/1910.
26. *Lanai.* Ka Nuhou, 9/9/1873.
27. *Laukiekie.* Ka Leo o ka Lahui, 6/22/1894.
28. *Lawai'a, mahiai, a me kalepa.* Ka Nupepa Kuokoa, 4/30/1925; 7/2/1926.
29. *Legend of Hainakolo.* Hawaii Holomua, 5/16/1914.
30. *Makaikai ana a puni ka honuo.* Ka Nupepa Kuokoa, 11/7/1868.
31. *Makalei.* Hoku o Hawaii, 5/1/1928.
32. *Make emo'iole.* Ka Hae Hawaii, 5/1/1928.
33. *Moanalua i kela au i hala aku a o Moanalua i keia au . . .* Ka Nupepa Kuokoa, 3/3/1922.
34. Mokumaia, J. K. *Moanalua past and present.* Ka Nupepa Kuokoa, 8/31/1922.
35. *Moolelo kahiko o Hawaii.* Ka Hoku o Hawaii, 9/9/1929.
36. *Na he'e kulua.* Ka Lahui Hawaii, 10/14/1899.
37. *Na i'a a kakou e ai nei.* Ka Lahui Hawaii, 8/10/1899; 8/26/1899.
38. *Na kuhinia o ka Koolau.* Ka Nupepa Kuokoa, 10/26/1917.
39. *Na wahi pana o Ewa.* Ka Loea Kalaiaina, 7/22/1899; 10/7/1899.
40. *Namakapao'o.* Ka Nupepa Kuokoa, 3/9/1912.
41. *Noted places on Lanai.* Ka Nupepa Kuokoa, 5/31/1912.
42. *Puakaohelo.* Ka Nupepa Kuokoa, 12/23/1893.
43. *Saving the fish.* Ka Nupepa Kuokoa, 3/8/1923.
44. *A short trip to the Koolau.* Ka Nupepa Kuokoa, 9/25/1896.
45. *Some things eaten when the sea is rough.* Ka Nupepa Kuokoa, 3/17/1922.
46. *Story of Pamano.* Ka Hoku o Hawaii, 3/8/1917.
47. *Story of Punia.* Ka Hoku o Hawaii, 2/13/1913.
48. *Wahine kino uhane.* Ka Nupepa Kuokoa, 11/25/1909.
49. *Worship to bring fish.* Ka Nupepa Kuokoa, 11/24/1866.
50. *Some very ancient things pertaining to Hawaii.* Ka Elele Hawaii, 2/10/1846.
51. *Waka keaka i ka nai.* Ka Loea Kalaiaina, 6/17/1899.

75a RANDALL, JOHN E. Personal communication. 1970.
76. RICE, W. H. *Hawaiian legends.* Bernice P. Bishop Museum, Bull. 3, 1923.
77. SEALE, Alvin. *Some poisonous Philippine fishes.* Phil. Jour. Sci., 7, Sect. D, pp. 289-291, 1912.

78. STEINBACH, Erwin. *Bericht über die Gesundheitsverhaltnisse der Eingeborenen der Marshall-Inseln im Jahre 1893-94 und Bermerkung uber Fischgift.* Mitt. aus den deutschen Schutzgebieten, 8 (2), 1895. Reprint of 15 pp.

79. STOKES, J. F. G. *Fish poisoning in the Hawaiian islands.* Bernice P. Bishop Museum, Occ. Papers, 7 (10) : 210-233, 1921.

79a STOKES, J. F. G. *Notes on Kahoolawe.* Ms. in Bishop Museum.

80. STOKES, J. F. G. *Walled fish traps in Pearl Harbour.* Bernice P. Bishop Mus., Occ. Papers, 4 (3) : 23-36, 1909.

81. TEMMINCK, C. J. & SCHLEGEL, Hermann. "Pisces," in Siebold, P. F. von, *Fauna Japonica.* Lugduni Batavorum, 1842-1850.

81a TESTER, ALBERT L. Cooperative Shark Research and Control Program, Final Report, 1967–1969. University of Hawaii, Department of Zoology.

82. THOMPSON, J. W. (Labels on fish casts in Bernice P. Bishop Museum.)

83. TITCOMB, Margaret. *The mysterious case of the poisoned fish.* Paradise of the Pacific, Vol. 57, February, pp. 27-29; March, pp. 21-23; April, pp. 29-31, 1945.

84. THRUM, T. G. *Variety of fish in Honolulu markets.* Haw. Ann. 1897, p. 48; 1898, p. 50; 1899, p. 54; 1900, pp. 45-46.

85. VANCOUVER, George. *A voyage of discovery to the North Pacific Ocean and round the world . . . 1790-1795.* Vols. 1-3, London, 1798.

86. WALFORD, L. A. *Marine game fishes of the Pacific coast . . .* Berkeley, 1937.

87. WATERHOUSE, H. *Deep sea fishing.* Haw. Ann., 1899: 104-106.

88. WETMORE, C. H. *Concerning Hawaiian fishes.* Haw. Ann., 1890: 90-97.

89. WHITLEY, G. P. *Poisonous and harmful fishes.* Council for Sci. and Ind. Research, Melbourne, Bull. 159, 1943.

89a WHITLEY, G. P. *The fishes of Australia, Part I: The sharks . . .* Sydney, 1940.

90. YUDKIN, W. H. *Tetrodon poisoning.* Bingham Oceanographic Collection, Bull. 9: 1-18, 1944.

INDEX

NOTE: Hawaiian names of fish will be found alphabetically in the text, pp. 56–163, and are entered here only if mentioned also in other parts of the text.